Agricultural Biotechnology: Prospects for the Third World

Edited by
John Farrington

Overseas Development Institute

Distributed by
Westview Press, Inc.
5500 Central Avenue
Boulder, Colorado 80301

British Library Cataloguing in Publication Data:

Farrington, John
Agricultural biotechnology: prospects for the third world
1. Developing countries. Agricultural industries.
Applications in biotechnology
I. Title
338.1'62'091724

ISBN 0 85003 119 2

Published by the Overseas Development Institute, Regent's College, Inner Circle, Regent's Park, London NW1 4NS

Front cover design: aspects of the tissue culture of cassava and sweet potato.

Printed and typeset by the Russell Press Ltd., Nottingham

ISBN 0-8133-1082-2 (Westview)

Contents

Notes on Contributors

John Farrington is a Research Fellow at the Overseas Development Institute. He coordinates the Agricultural Administration (Research and Extension) Network and is conducting research into the organisation and management of agricultural research for difficult farming conditions in developing countries.

Martin Greeley is fellow of the Institute of Development Studies, University of Sussex. He has a long-standing interest in rural technology development, including energy and crop storage in South Asia. He is a member of an IDS team preparing a major research proposal on agricultural biotechnology and the Third World. His recent work includes co-authorship of a report commissioned by Appropriate Technology International entitled: 'New Plant Biotechnologies and Rural Poverty in the Third World'.

Brian Mahy has been Head of the Pirbright laboratory, AFRC Institute for Animal Health, for four years. He is chief consultant on biotechnology to a UNDP-proposed Postgraduate Centre for Education in Immunological Biotechnology in India, and manages the several contracts for overseas work placed at Pirbight. Formerly, at the University of Cambridge (Wolfson College), he was first an Assistant Director in the Department of Pathology, and subsequently Head of the Virology Division. He is author of numerous scientific papers and books on virology, including *Virology: a Practical Approach* (IRL Press, 1985) and *The Biology of Negative Strand Viruses* (1987).

Sinclair Mantell is lecturer at the Unit for Advanced Propagation Systems at London University (Wye College). He is a consultant on plant tissue culture and biotechnology to several UK companies. His research interests focus particularly on tropical root crops, and include *in vitro* grafting, plant tissue culture methods, and cell and tissue differentiation. His books include: *Principles of Plant Biotechnology* (co-author; Blackwell, 1985) and *Biological Diversity and Genetic Resources: Techniques and Methods — Mass Propagation using Tissue Culture and Vegetative Methods* (co-editor; Commonwealth Science Council, 1987).

Preface and Acknowledgements

A particular attribute of biotechnology is the scope it offers for increasing the proportion of value added at the processing stage in food and other commodities. Indeed, new technologies in, for instance, fermentation and cell culture make it possible to transfer to the factory whole processes which hitherto have taken place in the field. The definition of agriculture used here is therefore particularly broad: it covers not only plant and animal products, but also factory-based processes related to these.

Chapters 2 and 3 are edited versions of a paper originally presented by Sinclair Mantell and Brian Mahy respectively at a seminar held at ODI in May 1988 entitled 'Agricultural Biotechnology and the Third World: Prospects and Policy Issues'. Some initial ideas for Chapters 1 and 4 were first presented by John Farrington in his background paper for the seminar, and by Martin Greeley and Steen Joffe in their paper for Appropriate Technology International.

The editor acknowledges other contributions from seminar participants which have helped to mould the argument presented here. The efforts of Kate Cumberland in typing from a difficult manuscript are also gratefully acknowledged.

1
The Issues

John Farrington and Martin Greeley

Introduction

The central question addressed in this book — the likely effects on less-developed countries (ldcs) of agricultural biotechnology — cannot be examined in isolation from broader trends in agricultural technology, production and trade.

Williams (1987) sees in the decade of the 1970s a 'new stage in the transformation of world agriculture' with, he argues, the institutionalisation of sustained food productivity increases not only in the OECD countries, but also in the larger Asian and Latin American countries. Strategies have been implemented which stimulate agricultural investment, provide incentives for producers and sustain science-driven technologies.

Among many ldcs in the 1970s, higher rates of agricultural production stimulated more rapid income-driven increases in demand for food than could be met by domestic production, so that food imports in aggregate rose from US$45bn in 1972 to US$165bn in 1980. Food imports fell after 1980 as ldc demand was depressed in the wake of lower income growth rates, and large international debt burdens under high interest rates. Continued farm support and, as some would maintain (Williams, 1987), institutionalised technological advance, stimulated further increases in OECD farm output, leading to a near-doubling of world cereal stocks between 1980 and 1987, a three-fold increase in world coarse grainstocks, and of beef stocks in the EEC, and a 30% reduction in aggregate food prices.

Arguing that '...neither protectionist trade nor domestic farm support policies can resolve agricultural problems brought about by fundamental changes in the application of technology-driven production and changing patterns of world markets...' Williams

(1987) advocates a world farm and trade development programme to raise food consumption in ldcs. Such a programme would require:

— temporary acceptance by major food exporters of current market shares, and a 'ceasefire' in the export subsidy war

— de-linking of farm income support in OECD countries from output-related measures

— promotion of income-generating farm policies in ldcs to stimulate food consumption, including a more growth-oriented pattern of international debt adjustment

— disposal of food surpluses in ways supportive of increased consumption and food security in ldcs without disruption of international markets.

Although they might appear idealistic, the relevance of these objectives to our theme is that, if they are not achieved, world trends of agricultural technology, production and trade will already be set to the disadvantage of ldcs, before we even begin to consider the potential impact of biotechnology.

Definitions

Problems of definition have long meant that: ' "biotechnology"..... embraces so extensive, expansive and diverse a spectrum of biological principles, phenomena, materials, organisms, reactions and transformations that no longer can it logically be considered in the singular as a collective or mass noun' (IDRC, 1985). This broad range of applications across several industrial sectors requires that 'biotechnology' be qualified according to the context in which it is used.

For present purposes, biotechnology has been defined as: 'the application of scientific and engineering principles to the processing of materials by biological agents to provide goods and services' (Bull et al, 1982). Some of the principal processes and products related to biotechnology are defined in the Glossary (Annex 1). It is important to emphasise that biotechnology transcends the techniques hitherto specific to crop or animal improvement. In both areas it exploits biochemical, molecular and cellular techniques for germplasm manipulation for improvement of the genetic stock and for better control of pests and disease. Additionally, the fact that 'bio'-techniques are being developed across a spectrum of activities — from the improvement of farm inputs to the processing of crops and by-products — means that biotechnology should be viewed in a food industry context, not merely in that of crop or animal production. The increasing capacity to shift food production from the farm to the factory reinforces the need for this wider perspective.

Biotechnology — distinguishing features

R & D at the cell and molecular level

Biotechnology differs from biological processes long-practised in the field (eg. plant breeding) or in the factory (eg. fermentation). Through its capacity for manipulating living organisms at the level of the cell or molecule, it allows a far wider gene pool to be drawn upon in breeding than that available hitherto in the specific pools of sexually compatible plants or animals. It has two broad components:

First, at the molecular level, 'genetic engineering' has a capacity to decode the genetic make-up of organisms, manipulate it by the insertion of new genes from similar or entirely different plants or animals, and generate new, different products. This potential — in many cases still theoretical — has caused much speculation regarding its possible applications.

Second, it is important to stress the contexts within which such interventions occur: they rely on numerous 'supporting' techniques to assist in understanding the genetic make-up of plants and animals, in monitoring the effects of interventions and, through tissue culture, in regenerating cells of the 'engineered' product into full-size plants or animals. These novel techniques aspects of biotechnology, which support genetic engineering also serve as an adjunct to conventional techniques. For instance, in plant breeding, even a single change to the genetic make-up of a plant (which is the most that is currently possible) requires incorporation of the plant(s) embodying this change into conventional breeding programmes to ensure that other desired characters have not been removed, and to enable removal of any undesirable characteristics which the intervention has introduced.

Linkages across sectors

Biotechnology is characterised by linkages across industrial sectors which are found only to a limited degree in earlier seed, agrochemical and mechanical technologies. They include linkages in *materials*, *processes* and *techniques*.

Materials

—The use of agricultural, animal and forestry biomass (as principal or by-products) in food processing, pharmaceutical, chemical and energy industries

— the reduction of agricultural biomass to intermediate products (eg. glucose, fructose, dextrins, lactose) by chemical or biological processes and the design of other processes in the same or different sectors to use one or more of these to provide such feedstocks as carbon. Frequently an important characteristic of such processes is their ability to draw on alternative intermediate products to supply the feedstocks they require, according to market conditions.

Processes:

— enzyme catalysis in starch, detergent and dairy industries, with an increasing market in such applications as reagents for laboratory analysis and clinical diagnosis

— fermentation in food, drink, pharmaceutical, sanitation and energy industries.

Techniques:

— genetic engineering, involving both the transfer of specific genes into a new plant or animal host, and the range of methods designed to understand genetic conformations, and to trace the impact of gene insertion

— techniques at the level of the cell for the fusion and culture of cells, for embryo rescue and for inter and intra-species hybridisation.

Economies of scale

The existence of cross-sectoral linkages in materials, processes and techniques suggests that large companies which have commercial interests in more than one sector will obtain higher output per unit of R & D input in biotechnology. Biotechnology R & D companies were initially small in the USA and Europe. In the USA, the progression has been from small companies initially supported by venture capital, to public quotation, and thence to absorption by multinational corporations having the financial assets and marketing capacity essential to the marketing of biotechnology products (Dembo et al., 1987). Many evolved from small university teams researching a particular aspect of biotechnology. Numerous firms survive as small, specialist enterprises, but economies of scale, particularly as processes and products approach commercialisation, together with the high cost of R & D (to clone a single gene costs at least $1m if the research runs smoothly; at least 20 man-years are required for isolating a commercially useful new enzyme; only one useful new antibiotic is found for every 20,000 strains of organism screened), the long lag between research outlay and revenue from product or process sales, and the uncertainty of outcome all appear to have exerted pressure towards increased size of firm. By the end of 1987, the number of US biotechnology firms had grown to 350 from 219 in 1983, but the aggregate number of scientists employed rose from 5,000 to 16,000, representing more than a doubling of average firm size. A closer targeting of research on marketable products also emerges: over 75% of staff were scientists in 1983, a proportion which had fallen to 50% by the end of 1987, whilst the proportion of marketing staff had doubled.[1]

The role of the private sector

Commercial companies are involved in agricultural biotechnology R & D to a far greater degree than in conventional agricultural research. For instance, whilst expenditure on conventional agricultural research in the USA in 1985 was split almost equally between private ($2.1bn) and public ($1.9bn) sectors, finance for agricultural biotechnology was over 50% higher in the private ($150 m) than in the public ($95m) sector. (Moses et al 1988).

Private sector interest is stimulated by awareness of the long-term potential of the genetic engineering technologies being developed, and the prospect of patenting the outcome of R & D, be it process or product. Additionally, certain companies, particularly those producing agrochemicals, see important complementarities emerging between their own products and those generated by biotechnology. The engineering of plants for resistance to herbicides provides an important example and is discussed below. Most agrochemical companies have therefore acquired in-house expertise, principally through take-over of one or more of the small 'start-up' companies, and these are committing annual operating budgets to biotechnology R & D some 3 times greater than average.[2]

Several observers stress the global character of the biotechnology industry. Dembo et al (1987), for instance, see in the actions of multinational corporations (MNCs) steps which are intended to secure a dominant role in technology generation and in world sales of biotechnology products through:

— preferential access to research results of potential commercial significance, passing as much risk as possible over to others, typically to the public sector

— use of patent law to protect their interests

— take over of small but potentially competing firms initially financed through venture capital

— erosion of public sector capacity by commissioning research from universities

— seeking support from governments both in legislation (on patents, health and safety etc.) and in contracts for research, particularly on military applications of biotechnology.

With very few exceptions domestic markets in less-developed countries (ldcs) are too small and fragmented to be of interest to the MNCs that are leading its commercial application in industrialised countries. MNC interest in ldcs has therefore focussed on a small number of export products. For domestic products, ldcs are dependent on either local private sector research and development (R & D) or on publicly-funded R & D. In practice, the latter has proven, and is likely to continue to be the more important, and is considered in detail below. Some of the constraints faced by local private sector R & D are illustrated by reference to experience in (i) Brazil and (ii) the Philippines:

(i) By 1983 there were already 600 research scientists in Brazil involved in biotechnology-related activities (across health, veterinary, agricultural and food industries). Joint ventures were formed between commercial seed companies and university-based research groups, and other work (eg. tobacco) was funded by multinationals. Commercial companies are also involved in the production of plant hormones, inoculants, and vaccines produced by fermentation technology. Government policy, led by the Special Secretaryship for Biotechnology within the Ministry for Science and Technology, encourages private/public collaboration in an effort to promote industrial development of biotechnologies, but there are substantial obstacles to progress in this direction, including:

— the Secretaryship's inability to control the majority of research institutes since these fall under other ministries

— the exclusion of seeds from patent arrangement

— fluctuations in government support for research centres,

including politically-motivated changes in staff

— the difficulty of importing laboratory equipment and materials given the pressures from a protectionist domestic equipment industry

(ii) In the Philippines, the Philippine Agriculture Biotechnology Company, established in 1985, had collaborated with the University of the Philippines at Los Banos to develop a pelleted symbiotic fungus (ectomycorrhiza) for processing waste on-farm. Substantial problems were encountered, including: delays in registering the product for tax and duty purposes (as with many biotechnology products, it is not included in current customs classifications); subsequent failure to capture the export markets needed to take advantages of economies of scale in production: simultaneous withdrawal of the main candidate export market (Thailand) as a result of pressures from its 'green' lobby.[3]

Markets for biotechnology products

As Hacking (1986, p.256) notes: 'Predictions of markets for biotechnology products have two outstanding characteristics: they are expressed in tens of billions of dollars, and there is a variation according to source'. He goes on to quote world market estimates for biotechnology products across all industrial sectors varying from $40bn to $50bn in 2000. Part of the confusion arises from inadequate descriptions of what is being measured. It is not stated, for instance, whether total estimated sales include the bulk materials such as ethanol or citric acid in which biotechnology processes have been used, or whether they refer to the products of eg. recombinant DNA and monoclonal antibodies alone.

Despite uncertainties surrounding the size of global markets, certain estimates made in the early-1980s of the US market size for pharmaceutical products and techniques then being developed have proven close to the revised and more accurate estimates now available, and in some cases have proven to be underestimates. Thus, diagnostic kits were estimated to have a $500m US market by 1990; later estimates put this at $1.3bn[4] Comparable figures for monoclonal antibodies initially suggested a 1990 US market of $1bn, which has subsequently been revised to $1.1bn[5] Total US pharmaceutical markets for biotechnology products are estimated at $5.5bn by 1990,[6] against an agricultural market of $25m in 1988, expected to rise to $0.7bn by 1993.[7] World markets for biotechnology products in 1986 totalled some $205bn, the major components being (approximately) 40% antibiotics, 20% ethanol, and 10% high fructose syrups. Apart from continued expansion of pharmaceutical biotechnology products by 2000, a major expansion is foreseen in agricultural applications and in replacement of petrochemicals by biochemical products as the prices of agricultural products decline relative to those of hydrocarbons. R-DNA is seen as the enabling technology of widest application in these areas, and a market for its products of $40bn is estimated for 2000 (Hacking, op cit. p.257). By comparison, present global sales of crude oil are worth some $450bn per year.

The regulatory environment

Regulatory issues fall into 3 broad types:

(i) Public concern over the potential environmental impact of technological change in the agriculture and food industries is not new: legislation governing, in particular, the use of agrochemicals and of food additives has been widely strengthened in the last 2 decades. Public concern in the case of biotechnology is, however, heightened by the current lack of knowledge of the way in which genetically engineered organisms might behave once released into the environment. The debate concerns both environmental degradation and ecological imbalances arising from the use of biotechnology products and processes. The Dag Hammarskjold Seminar, for instance, addressed these issues at Bogve, France, in 1987 and pressed for an international biotechnology policy agreement to meet the needs of 'the majority of the world's people....while working in harmony with the environment'.

(ii) Continued investment by the private sector in biotechnology R & D depends *inter alia* on an environment in which patents on products and processes can be obtained and enforced. Whether living organisms should be patentable has been the subject of heated debate at both ethical and practical levels, as has its implicit consequence of a reduction in the flow of scientific knowledge into the public domain (and therefore becoming accessible to organisations working in and for ldcs).

(iii) There has arisen the question, particularly in the USA, of whether the unrestricted commercial availability of ccrtain types of biotechnology threatens military and strategic policy objective.

These issues are discussed in more detail below:

1. Health and safety

In the ***USA***, the President's Office of Science and Technology Policy (OSTP) has been seeking to develop coherent guidelines and regulations on biotechnology since the early 1980s, latterly (since late 1985) delegating this function to a Biotechnology Science Coordinating Committee (BSCC), constituted for an initial period of 2 years. This followed the publication by the OSTP of a *Federal Register* notice on 31 December 1984 intended to regulate the use of biotechnology materials outside the laboratory, but which contained numerous ambiguities. Chaired by an Assistant Director of the National Science Foundation (NSF), the BSCC has members from the Food and Drug Administration (FDA), the Environmental Protection Agency, the US Department of Agriculture (USAID), the National Institutes of Health, and the NSF itself.

The 'deliberate release' of genetically engineered organisms (ie. for field trials) is the most contentious regulatory issue. The Environmental Protection Agency (EPA) in 1987 approved three biotechnology trials — two designed to prevent frost damage, to strawberries and potatoes, which had already been approved by the regulatory system in the state in which they were to be tested, and one involving a bacteria-borne genetic marker system. In the light of its 1987 experience, the EPA has declared it unnecessary for small, low-risk trials to be subject to Federal review, and plans to vest universities and corporations with authority to conduct safety reviews of proposed trials through 'environmental biosafety committees'. Even if these committees gain public acceptance, the EPA is left with two problems: one of deciding what type of trial still requires Federal review, and that of deciding to what products the Toxic Substances Control Act should be applied. Problems of definition of key terms have proven particularly severe: 'biotechnology has little of the 'uniformity' needed to make governmental regulatory activity an easy process' (Klausner & Fox, 1988). Examples include the efforts of the National Institutes of Health's Recombinant DNA Advisory Committee to determine containment levels deemed necessary for handling any particular microorganism, and to define the levels of review which different types of trial involving r-DNA require.[8] Further examples include the lengthy EPA deliberations to reach legally-enforceable definitions of 'pathogen',[9] and of 'environmental release' itself.[10]

Pressure on the EPA from industry for an early settlement of the confusion is being countered by threatened legislation from environmental groups on a case-by-case basis, and the EPA senses the threat from individual states to establish their own patchwork of regulations, which could be prompted either by increasing impatience with the EPA's deliberations or by concern that environmental issues are being neglected. The fact that two consultative reports on biotechnology and environmental safety (from the National Academy of Sciences and the Office of Technology Assessment) take differing views on the extent of risk involved merely adds to the confusion, as did an abortive attempt in September 1987 by the EPA to survey the biotechnology industry's generation and treatment of solid wastes, which required much the same evidence as was already being supplied to the Food and Drug Administration.

The US Department of Agriculture's proposals to deal with 'deliberate release' proposals through its Animal and Plant Health Inspection Service appear less confused. Researchers will be allowed to seek exemption from review for organisms generally regarded as safe whereas those less familiar (or suspected of being less innocuous) will be subject to stringent review. However, the respective roles of USDA and FDA remain undefined in certain important issues, such as, for instance, whether a foreign gene that produces a pesticide protein in tomato plants should also be regarded as a food additive. With efforts to produce genetically-engineered versions of live vaccinia virus-based vaccines for AIDS, and the news that health care and laboratory workers have recently been infected with AIDS in the workplace, the attention of regulatory agencies (ie. the Occupational Safety and Health Administration) has been drawn particularly to safety in the fermentation industry.

In ***Britain*** the release of genetically-engineered organisms is regulated by the Advisory Committee on Genetic Manipulation (ACGM), which reports to the Health and Safety Commission. Three field trials were approved in 1987, after some modification to their design. One, at Rothamsted, involved a new potato resulting from the fusion of cells of a domestic variety with those of a wild South American species resistant to leaf roll virus. The other potato trial involves genetically engineered incorporation of selection markers and protein-producing bacteria. The third trial involves release of a genetically-marked strain of *Rhizobium*, the nitrogen-fixing bacterium. To guard against unplanned spread of the novel genetic information contained in these trials, they were modified at the recommendation of the ACGM to incorporate deflowering of the potato plants and manual weeding and harvesting. The ACGM's terms of reference embrace not only novel combinations of genes, but all 'organisms constructed by techniques that involve the exchange of genetic information between species'. The ACGM has decided to avoid the evident pitfalls of US efforts to enforce a pre-defined set of regulations by considering each submission on a case-by-case approach. It remains a consultative body insofar as there is no compulsion to notify the ACGM of 'deliberate release' proposals, nor to follow its recommendations, but researchers generally see it as in their interests to do so. Notification of r-DNA experiments is, however, compulsory. Prior to the 1987 experiments the ACGM had considered only one earlier (1986) 'deliberate release', designed to test the efficiency of viruses already used in biological pest control.[11]

'The biotechnology industry in Europe sees the issue of deliberate release in simple terms: it wants to be able to test and eventually sell genetically manipulated plants and bacteria. And it recognises that it is impossible categorically to rule out that any biological risk is involved. Therefore it is prepared for, and might even welcome, sensible regulations that apply throughout the countries of Europe. And yet there is already a divergence that ranges from Italy's lack of *any* official regulation to Denmark's complete prohibition — a divergence that shows little sign of being curtailed by attempts within the European Commission to harmonise the regulation'.[12]

Joint ***European*** action so far has included a series of tests in mid-1987 designed to assess the possible risks of deliberate bacterial release, carried out in the UK, France and Germany. Whilst the tests passed without incident in the UK, and with only moderate protest from ecology movements in France, in Germany they were met with substantial 'Green' resistance, and the West German parliament has been presented with a parliamentary commission report which recommended a 5-year moratorium on deliberate release. At present, proposals for publicly-funded field trials must be cleared through the Central Committee for Biological Safety of the Federal Health Office, whereas for industry this procedure is voluntary.

In ***France***, advice on deliberate release experiments is offered by a Ministry of Agriculture committee established in 1987, but submission of proposals is not mandatory, unless they involve the use of r-DNA technology, in which case they must be considered by a committee of the Ministry of Research and Higher Education. Some ten field tests of genetically manipulated plants were carried out in 1987, several of which involved trials on tobacco and potato made resistant to herbicide. Efforts to reach a coordinated EEC approach to 'deliberate release' have had only limited success. The European Biotechnology Coordination Group's 'Biotechnology Risk Assessment Task Force' reported in March 1987, proposing the creation of a European organisation for risk assessment, but no action has been taken from Brussels. A further initiative, taken by the Biotechnology Unit of the UK Department of Trade and Industry has offered to share with industry the costs of a programme of research on Planned Release of Selected and Manipulated Organisms (PROSAMO), aiming to develop methods for assessing the spread of genes under field conditions. Eight major European companies have expressed interest in participating in PROSAMO with a view to providing a technical base for policy formulation in the EEC.

Within the EEC bureaucracy, a directive on planned release was to be published in mid-1987, but has run into widespread opposition from member countries, and has been the subject of much internal wrangling among the Directorates General. Primary responsibility for drafting the directive has, for instance, been given to the environment DG which approaches the issues more from the angle of disaster avoidance than from that of risk assessment. Recent versions of the draft directive still involve cumbersome liaison procedures between EEC headquarters and member governments over proposed trials, with time lags that could easily add a year to the procedure for approval. Furthermore, there would be nothing to prevent any member state from adopting regulations more stringent than those contained in the directive. Legislation in Denmark, in the form of the June 1986 Environment and Gene Technology Act, is already more restrictive than in other member countries. This Act prevents the deliberate release not only of any organism resulting from r-DNA technology, but also any organism whose production involves gene deletion or cell hybridisation.

Unless difficulties arising from inadequate coordination of regulatory approaches within the EEC can be overcome at the field trials stage, the problems of registering marketable biotechnology products in agriculture are likely to be such as hamper commercial establishment of the industry. (Newmark, 1987)

In summary, although much more knowledge is needed in this area, it seems that in the next decade most types of intervention designed for plants and animals are unlikely to be environmentally threatening.

Several will involve genes which are naturally occurring (albeit in a different host). Genes are *removed* rather than foreign ones added, or are being inserted into hosts unlikely to have vigorous crossing capacity either within or across species. Permission has now been given in the USA for field trials of 'low risk' genetically engineered organisms, and decentralised regulation of trials is in prospect. The situation in Europe is more confused: West Germany and the Scandinavian countries have severely restricted field trials, although several have been conducted in Britain, and Italy appears to have no restrictive legislation. Efforts are being made to design a coordinated European approach to field testing, but it is too early to evaluate them.

2. Patents and property rights

Patents

To develop certain biotechnologies — such as those involving the cloning of a gene — is expensive, and patents are seen by many commercial companies as a key means of recovering R & D expenditure.

The particular nature of biotechnology innovations makes orthodox patents difficult to apply and so requires some revision to the patent process, for instance:

i. Biotechnology processes and products are particularly difficult to describe in a manner which is both technically and legally unambiguous. Inadequate description can result either in refusal to issue a patent, and/or inability to enforce legal protection of a patent.

ii. Technical problems exist with the depositing and maintenance of biological materials — particularly the multicellular organisms — with the Patent Office.

iii. Much commercially valuable information does not fall within the realm of patent protection, including ideas, methods, laws of nature, the properties of matter and various kinds of business information, yet these can be critically important to the success of processes or products.

iv. 'Process' patents appear to offer particularly weak protection, especially where several alternative methods of manufacture exist, and the disclosure of information necessary to obtain the process described.

v. International cooperation in timing and disclosure requirements is weak so that infringement by companies located in other countries is difficult to prevent. This is particularly true of attempts to patent the 'processes' which currently in the USA form around half of the investment in biotechnology knowledge.

These difficulties have prompted two types of response among biotechnology companies:

(a) to seek modifications to patents legislation compatible with the characteristics of the industry

(b) to pursue courses alternative to patenting, such as trade secret protection.

Modifications to patents legislation. In the USA, biotechnology companies, through their two trade associations (the Industrial Biotechnology Association and the Association of Biotechnology Companies) have been seeking to strengthen patent legislation in two areas:

— extending process patent protection to products manufactured abroad. More than half of all US biotechnology patents are being granted for production processes rather than the products themselves. But the holders of such patents are not protected if the same processes are 'pirated', used abroad, and the products imported.

— restoring the patent term for agricultural products. Many agricultural biotechnology products are subject to extended pre-market regulatory review. The trade associations are seeking to lengthen the period of actual patent protection by making up for the lag between the granting of a patent and approval for commercial sale.

These issues are currently under consideration in the preparation of US amendments to US patent legislation. Other recent amendments include a ruling that animals incorporating biotechnology improvements are now patentable.[13]

Trade secret protection. Measures taken by commercial companies to strengthen their control over the processes and products resulting from their research, particularly in areas where patent protection is weak, include: confidentiality agreements with employees; strict in-house security procedures and editing of all published or verbal output; debriefing of departing staff to ensure that all confidential materials have been returned, and incorporation of a code into written output, or even proteins themselves, to facilitate identification of their origin. Providing that reasonable precautions have been taken to keep the results of research secret, a legal injunction can be taken to prevent any misuse of these by a competitor or former employee, and successful prosecution for misuse can result in both compensatory and punitive payments for damages (Payne, 1988).

Whilst trade secret protection is gaining ground as a complement to patents (Payne, 1988), the number of biotechnology submissions to patent offices continues to expand, rising in the USA from some 2200 in 1984 to over 3300 in 1986, with an approval rate of over 40%.

In Europe, much more progress has been achieved in unifying legislation on patents than on environmental protection. This began with the Paris Convention for the Protection of Industrial Property enacted in 1883 which required signatories to grant the same protection to non-nationals as to nationals of their own countries. The 1963 Strasbourg Convention led to harmonisation across European countries of definitions of 'new products' and agreement on how different a product must be in order to receive a patent. The 1970 Patent Cooperation Treaty provides a single international 'search' report to replace multiple national ones. By identifying relevant prior art, this report allows the applicant to abandon or modify his request before submission. The 1973 European Patent Convention (EPC) issues patents legally recognised in EEC member-countries, and because of the efficiency gains these offer over separate national registrations, there has been a substantial shift towards EPC and away from national patenting since 1973.

The EPC does not issue patents on 'plant or animal varieties or essentially biological processes for the production of plants or animals'. Microorganisms are an exception. Agreement has not yet been reached, however, on what constitutes 'essentially biological processes', and substantial divergence between EEC and national legislation exists in biotechnology products and processes, several countries (eg. France and Germany) allowing the possibility of patenting plant varieties that cannot otherwise be protected. Efforts to take the legal defences of EPC patents out of national courts and place them in a unified court system through the 1975 Luxembourg Patent Convention would, if realised, lead to further efficiency gains, but ratification is being threatened by political and constitutional objections raised by Denmark and Ireland. Recent reports (*Financial Times*, 5 October 1988, p 2) suggest that fear of emigration by

European biotechnology companies to take advantage of the more comprehensive patent protection offered in the USA and Japan have prompted renewed efforts to gain EC approval of new regulations granting protection on 'microbiological processes or products', plant varieties as such remaining under the protection of 'breeders' rights' at the national level.

Some progress has, however, been made in creating facilities specific to biotechnology. The 1977 Budapest Treaty on the International Recognition of the Deposit of Microorganisms for the purpose of Plant Protection deals with the deposit of living organisms in 13 internationally recognised facilities for purposes of patenting. Prior to this, applicants were faced with the cost and uncertain security of maintaining a culture deposit in each country where a patent was applied for. The pragmatic interpretation of the Treaty has been particularly useful in by-passing definitional problems: 'microbiological organisms' acceptable for patenting are those which it is biologically possible to maintain in the recognised depositories.

Since biotechnology products and processes are highly tradeable, commercial companies see substantial advantage in securing the widest possible patent coverage. However, they are faced by three types of problem:

— lack of enforceability of patents in countries which are not signatories to major conventions or have weak legal systems

— the varied quality of International Search Reports produced under the Patent Cooperation Treaty (PCT), (a problem arising from both language difficulties and from unfamiliarity with technical issues) so that several searches will have to be undertaken for countries as diverse as the USSR, Japan, France and the USA.

— the high penalties under some conventions (PCT; EPC) incurred by unsuccessful application for a patent. Applicants lose secrecy through the requirement to publish details of the product or process within 18 months of making the application, and lose rights to any deposit sample following abandonment or rejection of the application. Only in the USA, Japan and Canada are these penalties *not* incurred.

✓ *Property rights*

Of long-standing concern to ldcs, but heightened by the possibilities opened up by biotechnology, is the right of access to the genetic material indigenous to ldcs. These 'gene pools' contain the widest naturally occurring range of genetic diversity for many species of plants and animals, and have been used by both public and private sector based in the North as sources of material for breeding programmes, given that the North's own range of genetic diversity is limited both by natural conditions and by the introduction over wide areas of crops and animals having similar genetic configurations. Ldcs have begun to restrict access to these resources, conscious of the fact that North-based commercial companies, whether acquiring them directly or via such agencies as the International Board for Plant Genetic Resources (IBPGR), might use the material in products competitive with those exported from ldcs, and through patent legislation, may succeed in preventing ldcs from gaining access to the techniques and products concerned (Juma, 1988; Kloppenburg, 1988).

Given the apparent injustices which such possibilities involve, the defensive reaction of ldcs is not surprising. Emotive issues aside, however, it seems, as has been argued above, that North-based commercial concerns are likely to be interested only in certain ldc export products, that the possibilities of enforcing patents on these are limited, and (as is argued below) the possibilities of producing factory substitutes for these products through biotechnology are, with a few exceptions, technically and economically remote. Once these factors are discounted, the central question — and one not easily addressed by conventional neoclassical economics — is whether certain wild plant types are being permanently depleted or destroyed. The difficulties of establishing whether this is so, and what the economic value of such loss might be, need to be addressed if the debate is to be placed on a more reasoned footing.

On a more general issue, the impact that patenting may have on the supply of genetic material to ldcs, and on public access to technical knowledge has been a source of much concern in certain quarters (see, for instance, Hobbelink, 1987). Whilst in principle this concern appears valid, in practice, the public interest in general and that of ldcs in particular seems unlikely to be affected in the next two decades, if at all, by patenting, since:

i. there is healthy competition among biotechnology R & D companies
ii. many of the agricultural biotechnologies currently under investigation will not become commercially available for 10-20 years

iii. the specific technologies being patented (eg. herbicide resistance) are of little commercial relevance to ldcs

iv. public funding remains important in biotechnology R & D: 40% comes from the public sector in the USA, and virtually all the biotechnology R & D in and for ldcs is publicly-funded. These resources are sufficient to ensure an adequate flow of implementable biotechnology to ldcs, by a combination of fundamental research, monitoring of developments in the private sector and adaptation to suit the requirements of ldcs

v. patents will be virtually unenforceable in many ldcs.

3. Strategic issues

The US Department of Defense regards some biotechnology products and processes (eg. bio-reactors) as dual purpose given their potential for both civilian and military applications. Biotechnology products can therefore be exported only to a restricted list of countries (in practice, these are the same 21 — mainly industrialised market economies — to which the export of 'drugs and biologics' is permitted). The Defense Department has also sought to restrict the types of product or process to be exported. If these have both military and civilian significance, they are registered on a Military Critical Technologies List and the Defense Department has the authority to veto export of listed items, depending on such matters as how widely available they already are abroad (Gibbs, 1987). A forum for international agreement between the USA and its allies over the types of technology to be shared with potentially hostile nations is provided by a coordinating committee (CoCom) which meets periodically in Paris.

Recent reports[14] suggest that two levels of conflict exist regarding export policy: between initiatives towards trade restriction taken by the USA in CoCom and the more liberal approaches of its allies, and between the US Department of Defense and the Commerce Department, the latter broadly favouring promotion of biotechnology exports. However, even within the Commerce Department views on the extent to which exports should be encouraged appear widely divergent. For instance, its Biotechnology Advisory Committee has recently complained that its advice on ways to relieve the delays incurred by review and registration procedures have been ignored by officials within the Commerce Department. In brief, as with patenting and health and safety, administrative procedures are cumbersome, but it is difficult to improve these when administrators' technical knowledge of the issues is limited. There seems little reason to support the view that foreign trade in biotechnology products and processes will be restricted on strategic grounds.

Gene banks and genetic diversity

Optimistic views of new PBT claim that the prospects of creating new genetic traits through such techniques as r-DNA, irradiation and somaclonal variation make it unnecessary to maintain genetic diversity either in the field or in gene banks. By contrast, those taking a pessimistic view see in genetic erosion (such as that in south-east Asia where a single rice variety (IR 36) now covers 60% of the rice area, having replaced several thousand traditional landraces) an increased vulnerability to pest and disease epidemics such as the stem rust which caused substantial damage to the USA and Canadian wheat crop in 1950-54. Reality is more complex for several reasons: first, scientists' manipulation of existing gene conformations, or insertion of new gene types needs to be guided by evidence linking traits in existing host plants to its genetic identity; second, with many crops and for many traits the prospects of introducing useful 'foreign' genes remain so poor that naturally-occurring germplasm still provides breeders with a more useful set of building blocks.

The need to maintain as wide a range of diversity as possible is widely accepted, but it does not follow that the entire range of crops and varieties has to be grown in the field. Where agro-ecological conditions vary widely, a high degree of diversity *will* be maintained in the field, since a small number of varieties is unlikely to perform well under all conditions. Where conditions are more homogeneous, the risks of growing only one or a few varieties over a wide area are now much reduced for three broad reason: breeders have become aware of the need to incorporate into new varieties resistance to pests and diseases that does not rely on a single gene and is therefore not prone to sudden breakdown; second, not only is a wide range of genetic material now held at the IPBGR and other International Agricultural Research Centres (IARCs), but the characteristics of this material have been classified so that new sources of resistance can rapidly be bred into commercial crops where some breakdown is threatened. Additionally, new biotechnologies are increasing the efficiency with which gene banks can be maintained through, for instance, removal of duplicates (see eg. CIP, 1987). Awareness is also growing of the need to maintain germplasm across a wide front

— eg. in microbes (Kirsop, 1987). Third, new PBT are permitting expansion of the range of pest and disease resistant germplasm available in gene banks through such techniques as protoplast fusion which permit crosses between e.g. wild and cultivated varieties which are sexually incompatible and therefore impossible to cross with conventional techniques.

Notes

1. *Bio/Technology*, Vol 6, March 1988, p 244
2. *Bio/Technology*, Vol 5, February 1987, p 131
3. John Meadley, personal comment
4. *Bio/Technology*, January 1987, p 27
5. *Ibid.*
6. *Ibid.*
7. *Bio/Technology*, Vol 6, March 1988, p 243
8. *Bio/Technology*, Vol 5, February 1987, p 112
9. *Bio/Technology*, Vol 5, May 1987, p 424
10. *Bio/Technology*, Vol 5, March 1987, p 232; *Ibid*, April 1987, p 318
11. *Bio/Technology*, Vol 5, July 1987, p 675
12. *Bio/Technology*, Vol 5, December 1987, p 1281
13. *Bio/Technology*, Vol 5, June 1987, p 544
14. *Bio/Technology*, Vol 5, November 1987, p 1120

2
Recent Advances in Plant Biotechnology for Third World Countries

Sinclair Mantell

Introduction

If one accepts the term 'plant biotechnology' as the '*integrated* use of biochemistry, microbiology and chemical engineering to exploit plant materials and genetic resources for the production of specific products and services', then the Third World with all its natural and human resources, collected under the best solar radiation conditions known on this planet, should be in an excellent situation to make the most of the opportunities which recent advances in biotechnology now present. However, such are the complex historical, socio-economic, political and biological issues of ldcs that unless careful consideration is given to how the new technologies can best be exploited in the cause of development, a potentially significant opportunity might be squandered. To prevent such an undesirable outcome, it is first necessary to consider what are the recent advances which have a place in development, second to consider the realistic chances, at this comparatively early stage, of how much impact the various new technologies have already or are likely to make in the short term (the present to 1995) and third to consider how governments and development agencies might best *integrate* the new technologies into current and planned projects earmarked for the development of natural resources in the Third World. This paper argues that it is likely that appropriate *intermediate* biotechnology is likely to make the greatest impacts in Third World countries in the next decade as far as the development of crops and natural resources is concerned.

The Technologies

Several of the recent advances in plant biotechnology are of direct relevance to the developing world particularly for activities like crop improvement, forestry, environment stabilisation and for the future economic development of plant resources that have the recognised potential to provide a wider range of novel sources of industrial compounds. The latter can be used to produce a wide array of pharmaceuticals, agrochemicals, flavourants, enzymes, and polymers. As such, these precursors of core processes in the pharmaceutical, food, energy and the diagnostic industries represent strategic renewable resources for the ldcs. A summary of the most significant applications of plant biotechnology for the ldcs is presented in Tables 1a, 1b and 1c.

The recent advances are underpinned by research on laboratory techniques in the fundamental and applied plant sciences, plant tissue culture, molecular biology and biochemistry that interface directly with microbiology, agriculture and forestry. This makes them powerful tools for integration into conventional approaches to propagation, disease control, breeding, and the exploitation of natural resources. The special value that some of these advances have for the Third World is the fact that more can be done in a given time with quite basic resources. Thus, production of elite planting stocks of vegetatively propagated crops can be achieved in a comparatively short time as can the breeding of new crop varieties designed eg. to raise yields. Many of the techniques can often be developed in a small laboratory operation and integrated into existing plant improvement practices in order to achieve the same or enhanced productivity with less land and labour. Certain of the technologies such as micropropagation and meristem tip culture have been in regular use in the scientific research laboratories of developed countries for a substantial time — 30 years in some cases — while others such as rDNA technology and genetic transformation ('engineering') of crop plants are some of the more recent breakthroughs in plant science (Mantell, Matthews & McKee, 1985). There is still a great deal of field-based testing to be done on the first generation of transformed crops like tobacco, tomato and potato (1986 Yearbook of Agriculture). Furthermore, until the environmental issues are settled on an international scale for the widespread release of genetically engineered organisms into natural ecosystems, it is unlikely that, in the short term at least, genetic

Table 1a
Areas of recent advances in plant biotechnology relevant to the Third World

Crop Improvement	*Industrial Plant Products*
— rapid plant cloning	— novel pharmaceuticals
— micropropagules	— biomass conversions
— disease elimination	— fermentor technology
— germplasm storage/transfer	— factory production of phytochemicals
— haploid plant production	— diagnostics (virus and bacterial diseases)
— new avenues for hybridisation	— novel plant compounds
— *in vitro* selection	
— microbial innoculants	
— genetic transformation	

Table 1b
Improvement of agricultural, horticultural, forest tree crops and the environment using biotechnology approaches

Plant Propagation

* Production of disease-free, elite planting materials
* Rapid propagation and release of elite materials
* Transfer and storage of germplasm readily facilitated

Plant Breeding

* Reduction in breeding timetables: *in vitro* pollination, in vitro fertilisation, *in vitro* mutagenesis, *in vitro* selection
* Exposing somaclonal variability for increasing genetic diversity
* Selective addition and deletion of plant genes
  — somatic hybridisation, organelle enrichment and deletion
  — isolation and identification of specific plant genes
  — directed transfer of cloned genes using modified vectors

Plant/Microbial Interactions

* Development of effective mycorrhizal and bacterial inoculants
  — reduced fertiliser requirements
  — improved environmental stress tolerance
* Detoxification and recycling of biological wastes (composting)

Table 1c
Plant Biotechnology and Industrial Plant Products

Biomass for Renewable Energy and Material Sources

* Development of microbial fermentation processes to generate energy — biogas and gasohol
* Development of others to generate polymers of industrial value

Phytochemical Resource

* Identification of bioactive compounds particularly pharmaceuticals
* Characterisation of genes responsible for production of specific proteins in inert matrices for factory production of compounds
* Production of phytochemicals by plant cell cultures
* Development of biosensors and other diagnostic products

engineering will play a direct role in the production of novel crops suitable for use in tropical countries. This is not to say that the rDNA technology will not have its place in research laboratories of the Third World. On the contrary, India, Brazil, Mexico, Argentina and other countries are already planning projects involving rDNA approaches to crop improvement. Their policy appears to be sensibly conservative with the development of a few centres of excellence manned by scientists recently trained to Ph.D. level in Europe, USA, Japan and USSR.

The rDNA techniques now give plant physiologists, plant pathologists and plant biochemists new tools for determining how developmental events which occur during growth and in response to stress and disease situations are controlled at the molecular level. The information generated by this new activity will undoubtedly have long-term spin-offs for agriculture and its related activities. These are included in Table 2.

The specific technologies can be summarised as follows:

Rapid clonal propagation

The proven practical record of techniques such as micropropagation means that methods are quite well advanced and are being exploited for the aims of development in the ldcs. For instance, meristem culture has an established record of success in tuber and root crops like cassava, yams, sweet potato, Irish potato (for recent information, see Bryan, 1988) and fruit tree crops such as citrus. Despite the

Table 2
Some notable achievements relevant to the Third World

* production of disease-free planting materials
  — root and tuber crops, citrus, spices.
* rapid vegetative cloning of palms, citrus, eucalypts, tropical pines.
* international germplasm transfers using shoot cultures and micropropagules
  — potato, cassava, sweet potato, yams, dasheen.
* somaclonal variants in sugar cane, tobacco, potato
* somatic hybridisations leading to enhanced tolerance to stress
* recovery of salt-tolerant rice, wheat, sorghum, tobacco and alfalfa.
* herbicide-tolerant tobacco.
* identification and cloning of genes encoding various specific proteins.
  — acetohydroxyacid synthese (herbicide resistance)
  — papain and its related proteases
  — thaumatin
  — ricin
  — tuber storage proteins (potato, cassava, yam)
  — seed storage proteins (wheat, barley, maize, soybean, beans)
  — trypsin inhibitor (insect resistance)
* increased potential of microbial inoculants for increasing yields
* large scale fermentor production of plant metabolites — shikonin, ubiquinone.
* intermediate and high technology biomass conversion systems.
* Root Zone Bed (RZB) technology for water purification and environmental clean-up

successes, micropropagation systems are not yet commercially established in such major plant types as palms and tropical forest trees. This has recently been demonstrated by some setbacks in the large-scale cloning of elite oil palms in which a few of the clones produced have failed to give named flowering behaviour following a period of consistent vegetative growth. The preliminary status of the clonal palm industry is also highlighted by the fact that the main tissue culture production units pioneering this work on a large scale are based still in developed countries such as France, England and the USA. Although there are several smaller production laboratories based in ldcs such as Malaysia, India, Cote d'Ivoire and countries of North Africa, the cloning of palms by micropropagation will not be appropriately assessed on a large enough scale until field plantings of cloned palms are evaluated in detail for fruit yield and quality. This is not likely to be achieved until the early 1990's. Estimated world market capacity for oil palm alone is 100m clonal trees. Some of the larger micropropagation facilities in Europe such as Unifield Ltd. in Bedford, have current production capacities in the region of 1-2m oil palm plantlets per annum. The scale of plant production is such that further research into methods for planting out microplants through encapsulation of somatic embryos (see (iv) below) is now being explored seriously.

The fact that mature tree tissues are not generally amenable to vegetative proliferation is a technical hurdle for micropropagation. The integration of conventional stumping and hedging techniques with *in vitro* techniques such as micrografting to rejuvenate and reinvigorate mature tree tissues offers a solution to this problem (Bonga, 1982). The techniques of micrografting and the use of established micropropagated plants as mother stock have particular significance since these have been shown to lead to more responsive tissues for large scale vegetative propagation of some trees (Jonard, 1986; Howard & Marks, 1987). Somatic embryogenesis of tropical fruit trees such as mango, papaya and clove is being pioneered by US scientists in Florida and preliminary results are encouraging. However, as with the tropical pines, eucalyptus trees and palms, plant regeneration via auxiliary and adventitious shoots and/or somatic embryos requires the adequate field testing of *in vitro* produced materials before the widespread adoption of large scale cloning of tropical trees using new technologies can be achieved. Currently, there is evidence which suggests that precocious and/or abnormal flowering might be at least one side-effect of the tissue culture propagation of trees.

The main thrust in the application of micropropagation to forestry in the short term will be in the area of multiplication of pedigree seedling stocks to produce high quality seed orchards in forest species which can be propagated from either improved quality seed or from hedges of juvenile materials that provide conventional cuttings or explants for micropropagation. Such integrated procedures are beginning to find increasing use in some tropical forestry projects.

Micrografting

This technique involves the excision of mature meristems from fruit and forest trees and the grafting of these under aseptic conditions onto seedling rootstocks. The objectives of this method are to obtain disease-free scion materials of fruit trees, e.g. citrus (Navarro & Juarez, 1977) and to obtain rejuvenated shoots from mature trees (see above).

International germplasm storage and transfer

Aseptic shoot cultures of disease-indexed crop plants like Irish potato, sweet potato, cassava and yams are being used on a regular basis for the international exchange of germplasm. In so doing, the inadvertent transfer of soil-borne pests and diseases caused by fungi, nematodes and insects is circumvented. There are still some problems, however, with incipient bacterial infections which can often increase in frequency after airmail transfer following temperature stresses encountered during transit (Roger Bancroft, *pers.comm*). The uses of cryopreservation for germplasm storage of tropical crop germplasm have so far been demonstrated in a few crops like cassava and beans. International transfers of Irish potato, sweet potato, and cassava are now routinely carried out between tropical and temperate research and development centres in the CGIAR/University networks. This facilitates collaborative research between East/West and North/South in, for example, the disease indexing and quarantine of sweet potato and cassava distributed from CIAT to Africa (Roca, Rodriguez , Beltrau, Roca & Mafla, 1982).

Micropropagules

Miniature perennating organs such as tubers, corms and bulbils can be produced *in vitro* on shoot cultures of a range of vegetative propagated crops like Irish and sweet potato, yams and dasheen. These micropropagules make potentially important materials for germplasm transfers and several international germplasm projects are using this form of exchange system, e.g. potato transfers from CIP, in Peru and yam transfers from Wye College, UK. These propagules, akin to conventional forms of planting material, allow improved germplasm from *in vitro* based propagation and disease elimination programmes to be effectively distributed and integrated into ldc agriculture (Mantell & Hugo, 1986). Just as the minisett technology of yams is now being adopted in West Africa and the Caribbean, the microtuber concept could well best meet the requirements of getting mass produced pathogen-tested materials into the field. This stage of laboratory-to-field transfers is often the most critical in implementation of tissue culture technology since it takes special husbandry conditions to raise sturdy plants from the initially delicate microplants transferred to soil. Somatic embryos are now being embedded in desiccated gels or in fluid-drilling gels for investigating the prospects of delivering mass-produced somatic embryos, raised in fermenters, directly into the seed row (Fujii, Slade, Redenbaugh & Walker, 1987). This is becoming more established technology in greenhouse planting of carrot and alfalfa somatic embryos. It will only be a matter of time before 'artificial seeds' are introduced directly into the field using techniques similar in principle to fluid drilling and spray casting.

In vitro techniques useful for breeding

These approaches are at a relatively early stage of development in tropical countries but their value is now being recognised and their inclusion in conventional breeding schemes is now on the increase. It must be stressed that the *in vitro* techniques mentioned here supplement, not replace, conventional breeding strategies and procedures. They have great potential for shortening significantly the breeding cycles required for given levels of crop improvement.

Somaclonal variability

Somaclonal variants generated from callus cultures of sugar cane, tobacco, sorghum, potato, rice and wheat have provided breeders with new sources of variability to incorporate into conventional breeding programmes. Agronomically useful traits such as increased tolerance to physiological stress and pests and diseases have been recovered from such materials (Scowcroft, 1985). These genotypes could be of particular significance for the difficult farming areas prevalent in many parts of the Third World.

In vitro selection strategies

Heavy metal, salt and disease tolerance have been recovered in regenerated plants of alfalfa (McCoy, 1987), tobacco (Nabors, Gibbs, Bernstein & Meis, 1980), rice (Yano, Ogawa & Yamada, 1982), and many others (Chandler & Thorpe, 1986). In those cases where fertile plants have been recovered, new opportunities have been created for some of the germplasm generated to be included in breeding trials. The results of these efforts are only at a preliminary stage and until remaining anomalies in the expression of salt tolerance are better understood, an empirical approach to breeding will remain the only alternative.

Embryo rescue

The rescue of small immature embryos derived from zygotes

produced from wide crosses between wild and domesticated species is one short-cut breeding strategy for obtaining new variability which can be exploited relatively easily in crop improvement. Speedy improvements to the germplasm of tropical forage legumes, rice and maize are now being obtained in this way. One good example of this is the utilisation of genetic material (through wide crosses and recovery of hybrid embryos) from wild beans containing the protein arcelin which has conferred insect resistance on several *Phaseolus* field bean varieties (work done at CIAT, Colombia).

Haploid plant production

Relatively short periods of time are all that are needed to generate homozygous breeding lines of tropical cereals if haploid or dihaploid plants can be regenerated from immature pollen. Conventional pedigree methods of homozygous plant generation take 5 — 6 years and longer but in some cases regeneration of haploid plants has been achieved in less than two years, e.g. for barley, tobacco and rice. The method therefore is saving valuable time and field space.

Somatic hybridisation

Protoplast fusion represents one way to create much-needed gene-flow between wild species with stress tolerance features and intolerant cultivated species (Cocking, 1985). Provided regeneration of plants from isolated protoplasts is possible (and this is becoming more so as research is being carried out on tropical species by a greater number of laboratories in the North and in ldcs) then the feasibility for manipulating useful breeding traits such as male sterility by organelle enrichment and/or exclusion strategies will become a practical prospect. The recent reports of plant regeneration from isolated protoplasts of maize, millets, rice (Marx, 1987), tropical forage legumes point the way to a dramatic expansion in the breeding possibilities through somatic hybridisation. Laboratories in India, Pakistan and Brazil are particularly well placed to capitalise on these recent developments. Proven successes in the use of somatic hybridisation in the development of new breeding lines of tomato, potato and brassicas can spur scientists in Third World countries to do the same with indigenous crops.

Identification and cloning of plant genes

Before useful genes can be transferred from bacteria into plants or from plants into cloning vectors like bacteria or lambda phase and then transferred into crop plants in order to confer improvements to certain agronomically important traits, genes must first be identified, then isolated and cloned with appropriate promotor, intermediate and terminal sequences to facilitate their optimal expression in other organisms. This activity is well under way in the developing world and at least fifty biotechnologically significant plant genes have been characterised, sequenced in some cases and cloned. Their transfer into microbes and plants is also well-advanced in corporate research laboratories of commercial companies. Notable examples are various herbicide resistance genes and the seed storage protein genes of peas, beans, wheat, maize, barley and other legumes and cereals and tuber storage protein genes of potato. Insect resistance genes encoding a protein toxin of *Bacillus thuringiensis* active against *Lepidoptera* pests (caterpillars), and virus coat-protein genes of TMV that confer cross-protection to viral attack (Chilton, 1988) are significant genes for the improvement of crop plants which have recently been transferred into or between different plant species.

Other important groups of plant genes are those concerned with enzyme steps in nitrogen fixation, photosynthesis, fruit ripening, starch synthesis and disease resistance. All of these genes have great significance for the eventual improvement of crops grown in Third World countries but it is likely that the seeds for such genetically engineered crop plants will be produced under tight commercial control of the multinational seed companies. To recoup high R & D costs, such seed are likely to be priced beyond the reach of ldc farmers, but the reduced levels of agrochemical input they may lead to in the North may be environmentally positive. They may also have a stabilising effect on world food prices. One further outcome of genetic engineering activity may be various types of 'philanthropic' activities which could help in the genetic improvement of subsistence crops in ldcs. It is foreseeable that some of the MNCS involved in biotechnology research might seek to improve their corporate image by providing certain rDNA engineered molecules which could assist public-funded research aimed at genetic improvement of subsistence crops like cassava and yams. For instance, tuber storage proteins are encoded by genes which can be altered individually. Levels of proteins accumulating in basic subsistence food crops can therefore be manipulated to provide increased food value.

The transfer of cloned genes into crop plants can be achieved readily for some dicotyledonous plants by means of either the Ti or Ri plasmid-based systems (Horsch, Fry, Hoffman, Eichholtz, Rogers

& Fraley, 1985). There has recently been some success using these vector systems for genetic transformation of maize (Rhodes et al., 1988) and yam (Schaefer, Gorz & Kahl, 1987). The development of these systems for tropical cereals may come through the use of electroporation of microspore protoplasts (Schillito, Saul, Paszkowski, Mueller & Potrykus, 1985) or through the use of modified *Agrobacterium* systems using tandem copies of viral DNA to increase transformation frequencies e.g. the maize streak virus systems of 23.

Plant/microbial interactions

The more widespread use of organic farming systems in the ldcs could lead to significant impacts on the environment and the socio-economic structures of farming communities. Some recent advances in this field are not solely the results of advanced laboratory experiments.

Nitrogen fixation symbioses

Microbial inoculants have been routinely used in the ldcs wherever soybean is cultivated since its production depends on the presence of the symbiont *Brachyrhizobium japonicum* which has a restricted natural distribution. Therefore, all soya grown outside China is treated at seed sowing with artificially produced inoculant. This means that the inoculation production technology is becoming more widely available in Developing Countries. Few other crops are inoculated at present but recent advances in molecular biology of *Klebsiella* and *Rhizobia* — nitrogen fixing bacteria — means that improved strains of the bacteria may now be tailored to suit specific crop species. So far these advances have only produced some 10% increase in nodulation in leguminous plants with only minor consequent improvements to yields. However, recently an ingenious method of soil-core testing has facilitated the screening of effective inoculants which give highly significant increases in vegetative development and yield. Examples of such encouraging soil-core testing results are now known for *Rhizobium* on clover and on the tropical forage legumes *Centrosema* and *Kudzu* on acid soils in the tropics (Ken Giller and Rosemary S. Bradley, *pers.comm.*). Advances such as these could have important impacts on crop production in the Third World, leading to less reliance on inorganic fertilisers and to increases in the levels of production because of the creation of more stress-tolerant crop stands resulting from the increased availability of nutrients to plants.

Composting and recycling of biological materials

The upsurge in intermediate technoloy over the last decade in ldcs in South Asia, the Middle East and North Africa has resulted in many versions of on-farm recycling processes which have been extremely successful on a practical level for producing non-toxic organic fertilisers and energy sources like methane (biogas). Some recent advances in molecular biology and characterisation of thermophillic microbes which are active in such processes are leading to a greater understanding of degradation processes and major improvements to the intermediate technologies in current use can now be made. Increased technology has led to some extremely efficient composting processes in temperate countries. These need now to be tested under rural and urban situations in the Third World for producing environmentally safe organic manures.

The creation of ldc technological capacity

Following from this description of the 'state of the art', the principal areas where future technology development offers the highest returns to developing countries, and where effort should therefore be concentrated, are summarised in Table 3.

Table 3
Some current needs

* increased efficiency and diversity of micropropagation techniques
* regeneration of plants from protoplasts of major leguminous and cereal crops
  — rice
  — forage legumes, soybean
* more efficient haploid plant production in cereals
* more consistent somatic embryogenesis in several tropical fruit and forest trees
* gel encapsulation of cultured somatic embryos — oil palm and vegetables
* genetic transformation of maize and other monocotyledonous crop plants
* development of specific diagnostic tests for plant viruses using rDNA technology.

In a more general vein, the development of technological capacity requires a steady supply of funds, good managers, dedicated technicians and scientists working in concert with the recipients of the technology. The logical way to ensure that all these inputs are satisfactorily supplied is to strengthen support for link and network programmes between managers and scientists in universities, government and quasi-government institutions based in North/South or North/East/West countries. Industrial partnerships should be encouraged wherever appropriate to foster future investment in projects initiated by national funding agencies. International funding agencies should play a more ambitious role in providing support for the collaboration of research and development groups in biotechnology. These larger organisations should also continue to provide sources of information to the Third World on biotechnology matters.

To be realistic, policies should be guided by the currently achievable impact values of technologies, not on predicted impact values which may take considerable time to materialise. Therefore, I would urge that for the most effective development impact through plant biotechnology one should take three spikes to the fork:

1. Support should be given wherever possible for appropriate intermediate biotechnology which integrates well with existing activities in crop improvement and exploitation of natural resources (i.e. plant tissue culture, and forms of this technology which are relevant to *in vitro* and conventional breeding as in Table 3).

2. Support for technology of appropriate intermediate types which generates practical gains in integrated ecosystems (e.g. microbial inoculants and composting, biomass for industrial processes).

3. Wait and see what the rDNA technologies have to offer for development. In the meantime assist in the training of scientists of developing countries in collaborative R & D projects based in both developed and Third World countries. Identify 'centres of excellence' in various Third World countries where plant molecular biology competence is proven or most likely to generate impact due to good management. Support wherever possible European Molecular Biology Organisation-type student exchange schemes in the molecular sciences. It is the future human resource which is likely to be so critical to the success of rDNA in development of plant resources. It is also worth reflecting that

there are many development-conscious graduates in countries of the North eagerly waiting for their chance to contribute to the improvement of yam, cassava and winged bean rather than wheat, oil seed rape and evening primrose. The potential benefits of harnessing this enthusiasm amongst our own science graduates for the cause of development of Third World countries should not be underestimated.

3
Recent Advances in Animal Biotechnology for Third World Countries

Brian Mahy

Introduction

Animals are an important element of the agricultural economy of developed countries through provision of food, industrial raw materials (wool and hides), and manure. In developing countries an additional function of animals is the provision of draught power (Bodet, 1987). For example in India, some 60% of draught power is supplied by cattle. World production and utilisation of products of animal origin was recently reviewed by Blajan (1987). The relative values of world animal production in 1980 are given in Table 1:-

There is considerable scope for improving productivity in developing countries. This is made clearer when one considers the

Table 1
World Animal Production 1980

Gross Value, Millions of US dollars

Countries	*Meat, Milk, Eggs*	*Hides, Skins, Wool*	*Draught*	*Manure*	*Total*
Developed	196,000	6,000	6,000	4,000	212,000
Developing	92,000	4,000	40,000	6,000	142,000

(Source, *FAO Situation Report*, 1982)

Table 2
World Population, Resources and Meat Consumption

	Developing Countries	*Developed Countries*
Population	74%	26%
Plant Production	50%	50%
Animal Production	25%	75%
Meat Consumption (per capita)	13.5kg	75.4kg

(Source, FAO, *Production Yearbook*, 1985)

world population distribution and resources are considered in relation to food consumption (Table 2):

Hides and skins are economically important by-products of animal production, but the rate of slaughter is determined by the demand for meat, not the demand for processed leather and leather goods. In developed countries, skins are of good quality, but in developing countries defects caused by diseases and ticks greatly reduce their value.

There are other animal-based industries which are important to developing countries. Fish production has increased in recent years, and of the total world fish production of some 85 million tons, developing countries produce about 45%. Aquaculture contributes only 10% of total fish worldwide, and there is considerable scope for its development in Third World countries.

Honey production amounts to about 1m tons, with developing countries producing about 43% of world output. Finally, silk is an important animal product of high value. Some two-thirds of the world production of 60,000 tons of raw silk comes from Asia, and it is important to the economies of China, India, South Korea and Brazil.

Improvements from biotechnology

Biotechnology offers several new approaches to animal production and health which could greatly benefit Third World countries. The possibility of introducing these new technologies will depend largely upon education and the development of skills and of an appropriate supporting infrastructure, but there is no doubt that developing countries are aware of the range of new technologies available, and are anxious to participate in the implementation of biotechnology for agriculture. They will not be content to see biotechnology used and applied only by scientists from developed countries, but seek training in what they believe is a vitally important new science which will aid their agricultural economy. The main contributions from biotechnology in the near future would appear to be as follows:-

Table 3
Biotechnologies for Animal Production and Health

Technique	*Advantage*
Multiple ovulation and embryo transfer (MOET)	Accelerated cattle breeding and improvement. Ease of transport of genetic material between countries.
Transgenic animals	Introduction of new genetic traits, e.g. growth rate or disease resistance.
Monoclonal antibodies and cloned DNA probes	Improve disease dignosis.
Genetically engineered and Molecular Vaccines	Improved health care.

Multiple ovulation and embryo transfer

Multiple ovulation and embryo transfer can be used on any species, and has been successfully used for human medicine; the first baby resulting from this technique was born on 25th July 1978. In animals, more than 100,000 embryo transfers are now carried out annually in the USA, and some 30,000 in Europe. In principle, genetically superior animals are used as donors, and superovulation is induced by hormones such as gonadotrophin. After artificial insemination, embryos are recovered by flushing and may then be stored frozen indefinitely in liquid nitrogen before transfer to a recipient animal. Embryo splitting is possible, preferably before freezing, which can double the number of embryos available for transfer (Wagner, 1987).

Successful embryo transfer requires highly motivated and experienced staff, and a high capital investment for facilities, equipment and drugs. Nevertheless it should be possible to introduce the technique to less developed countries, especially through nucleus breeding units (Smith, 1988), with the prospect of much more rapid improvement of animal genetic stocks than is possible by conventional breeding programmes. However, set against this is the fact that most introduced 'highly bred' stock is highly susceptible to local indigenous diseases.

Transgenic animals

By contrast, production of transgenic animals is a highly sophisticated procedure which is still in its infancy so far as farm livestock are concerned.

Gene transfer provides new sources of genes, since genes may be transferred between any species, even where natural mating is impossible. To achieve permanent change in a strain of animals the newly inserted genes must be stably inherited. In theory, the ability to obtain and manipulate genes as cloned DNA copies enables the insertion of new genes into the germ line of animals but they must be in the appropriate configuration and location so that the phenotype is expressed in the right tissue at the correct stage of development.

The current techniques have been worked out in mice, which are first made to superovulate through hormonal treatment, then eggs are collected and the foreign DNA containing the new gene is microinjected into the larger male pronucleus of the fertilised one-cell egg. Some 10-20% of the eggs survive microinjection and develop in the oviducts of pseudopregnant foster mothers. Of the mice which are born, some 20% will carry one or more copies of the desired gene in their germline. Other techniques for gene insertion, involving retrovirus vectors or plasmids based on mammalian parvoviruses, are being explored.

The transfer of this technology to farm animals is difficult and expensive, but some success has been achieved in pigs and sheep. Nevertheless, the current success rates are at best of the order of 8% transgenic births in these species (Clark et al., 1987). The main problem in these species is to visualise the pronuclei for microinjection. In poultry, the application of micromanipulative techniques suitable for mammalian eggs is much more difficult, and instead the possible application of defective retroviruses, as vectors for gene insertion, is being explored (Freeman and Bumstead, 1987).

The creation of hybrid DNA molecules, in which the regulatory sequences from one gene are joined to the desired coding sequences of another gene, has the potential to achieve tissue-specific expression, governed by the regulatory sequences. This has been achieved in a number of cases with varying success (Brinster et al., 1988), but the technique is still somewhat unpredictable and difficult in many species of agricultural interest.

Gene transfer, when refined, may enable production of animals with desirable genetic characteristics such as faster growth rate, increased wool production, or resistance to specific diseases. Once such animals are produced, the maintenance of these characteristics could then be achieved by embryo transfer. Although an intense interest in transgenic animals exists in developing countries, the very high costs involved would seem to preclude transfer of this technology to less developed countries in the immediate future, although the introduction of newly created 'breeds' by embryo transfer should be possible.

Disease problems

Whatever the improvements possible in breeding, a major hurdle in their application to developing countries remains in the many diseases

Table 4
World Costs of Principal Diseases of Animals
(in millions of US dollars)

Species	*Disease*	*Millions US$*
Cattle, sheep, goats	Foot-and-Mouth	50,000
	Rift Valley Fever	7,500
Cattle	Shipping Fever	3,000
	Bluetongue	3,000
	Leptospirosis	4,500
	Brucellosis	3,500
	Mastitis	35,000
	Calf diarrhoea	1,750
Pigs	Aujeszky's	650
	Pasteurellosis	500
	Trichinellosis	2,500
	Gastroenteritis	1,800
Poultry	Retrovirus Infections	1,000
	Fowl cholera	200

(Source, *The impact of biotechnology on animal care*, Technology Management Group Inc., 1986)

Table 5
Worldwide Wastage from Animal Diseases
(millions of tons)

Commodity	*High-income countries*	*Developing countries*	*Total*
Beef cattle	44.6	4.6	49.2
Sheep and goats	0.6	1.2	1.8
Pigs	4.2	4.6	8.8
Poultry	2.2	1.9	4.1
Milk	53.9	30.5	84.4
Eggs	2.3	2.8	5.1
Total	107.8	45.6	153.4
Loss rate	17.5%	35.0%	

(Source, FAO)

which have yet to be controlled. In several instances, it has been shown that new breeds are even more susceptible to disease than indigenous breeds. Examples of the current costs of major animal diseases worldwide are given in Tables 4 and 5.

Two-thirds of the total wastage from animal diseases worldwide occurs in developed countries, because of the high productivity. But wastage through disease in developing countries, at 35%, continues to hinder their livestock production.

Biotechnology in disease control

Diagnosis

Successful control of disease requires recognition and accurate diagnosis. This has been greatly improved in recent years through developments in biotechnology. The most versatile has been the introduction of monoclonal antibodies combined with enzyme-linked immunosorbent assay (ELISA). Once appropriate monoclonal antibodies have been obtained, they can be supplied to developing countries directly or in the form of kits which use precoated plastic plates and simple reagents for completion of the tests (Ferris et al., 1988). As an example of this technology transfer, Pirbright laboratory is cooperating with the International Atomic Energy Agency in Vienna to supply African countries with kits for rinderpest virus diagnosis. In some cases, where monoclonal antibodies are unavailable, or not of the desired specificity, other techniques such as nucleic acid hybridization may be appropriate. One such test developed at Pirbright is used to distinguish infections caused by *peste des petits ruminants* virus (PPRV) from rinderpest. The symptoms of the two diseases are clinically indistinguishable, but although the causative viruses are distinct, they cannot be distinguished antigenically with available serological reagents.

Each virus possesses a single-stranded RNA genome, which was copied by reverse transcription into cDNA, and cloned in a bacterial plasmid. DNA clones of the nucleoprotein (N) gene from rinderpest and *peste des petits ruminants* viruses did not significantly cross-hybridise, and can therefore be used as probes to differentiate the infections (Diallo et al., 1988). Such techniques are important in tracing the origin of a disease outbreak. For example a recent rinderpest outbreak in Sri Lanka, the first such outbreak for forty years, was shown to be rinderpest and not *peste des petits ruminants*, although the disease was apparently introduced by infected goats from India.

Vaccines

Genetic engineering has made possible several new approaches to vaccines which are safe, effective, and specific. However remarkably few vaccines produced in this way are in current use. They include a hepatitis B vaccine from a genetic recombinant made in yeast, a vaccine against diarrhoea in neonatal pigs, which uses E.coli genetically engineered to express K88 and K99 antigens, and a vaccine against Aujeszky's disease in pigs which consists of an engineered gene deletion mutant of the virus.

Considerable potential exists for the introduction of vaccine viruses such as *vaccinia* which are engineered to express a foreign virus gene or genes (Moss et al., 1988; Tartaglia and Paoletti, 1988). *Vaccinia* has a considerable advantage of stability during storage and transport in freeze-dried form, and in the effectiveness of the induced immune response. However, *vaccinia* virus presents a significant hazard to man, and seems unlikely ever to be used.

Because of the human hazard, several laboratories are attempting to introduce genes into other pox viruses notably fowl pox (Binns et al., 1988; Boyle and Coupar, 1988) and sheep and goat pox (Gershon and Black, 1988).

Another approach to vaccines is to use simple synthetic peptides which represent antigenic sites or epitopes on the surface of the pathogen. The required amino-acid sequences for the peptides can be predicted from nucleic acid sequence analysis. So far, peptides against foot-and-mouth disease virus seem to offer the best potential for vaccine development (DiMarchi et al., 1986). Finally, the approach of making antibodies against defined monoclonal antibodies (anti-idiotypes) for use as protective vaccines for certain diseases is being explored. These offer potential with complex disease-inducing organisms such as parasites (Murray, 1987).

Transgenic animals

Perhaps the most exciting long-term aim is to produce disease-resistant species of farm livestock by manipulating the germ line. These might involve genes expressing antigens which become expressed on the cell surface and prevent entry of a pathogen, or genetic modification so as to delete receptors required for entry of an intracellular pathogen. At present the prospects for this approach appear promising but some way in the future. In addition, a problem with all transgenic animals will arise if regulatory bodies, as currently in the EEC, do not permit their use for human consumption (Glosser, 1988).

4
Potential Implications of Agricultural Biotechnology for the Third World

Martin Greeley and John Farrington

Future prospects

Hacking (1986, p.258) estimates biotechnology companies' breakdown of expenditure at 61% for human health care, 23% in agriculture and 16% in others (mainly chemical processes). Health care products have been, and will continue to be, favoured because there is a clearly-defined path from laboratory experimentation to market. The efficiency of, for instance, a protein, can be tested by established procedures, and its market accurately estimated. Obtaining regulatory approval is long, costly and risky, but approval is not required for diagnostic kits, such as those based on monoclonal antibodies (which have both medical and veterinary applications), nor is it as difficult to obtain for veterinary products, so that these appear to offer the most secure prospects for expansion over at least the next decade.

By contrast, plant genetics is a less well-defined area. Knowledge at the molecular level is more limited, vectors for gene insertion are fewer in number, and patents are difficult to police. Detailed discussion of future prospects for plant and animal applications is presented in Chapters 2 and 3 respectively. These are summarised below, as are the prospects in such other areas as industrial fermentation, plant cell cultures and forestry.

Plant breeding

Leading representatives of US biotechnology companies were asked at a recent round table on plant biotechnology[1] to predict where the industry would be in 5 and in 15 years' time. The consensus that emerged was that a number of synthetic seeds would be commercially available by 1992 in industrialised countries, including those conferring herbicide resistance in eg. tobacco, and pest resistance in tobacco, potato and tomato, through the incorporation of genes for *Bacillus thuringiensis*. R & D programmes would have identified, isolated and cloned some 100 genes of commercially useful plants, and field trials would be underway with roughly half of these. Some of these are likely to be of major commercial significance, such as those modifying soya plants to produce oil with the characteristics of sunflower oil.

The range of crops for which genetically-engineered seed could be available in 15 years' time was considered to be much wider — possibly including cereal crops which, until recent successful research, had proven difficult to engineer. Traits depending on changes not to one but to several genes (yield; tolerance of environmental conditions) might by then also be on the market. On the other hand, the genetic processes underlying some of the well-publicised possibilities such as the transfer of nitrogen fixing capacity to plants other than legumes are extremely complex, and must be regarded as longer-term. It is known, for instance, that there are at least 17 genes coding for the enzymes involved in the fixation of nitrogen by *Rhizobium*, the bacterium found in the nodules of leguminous plants. To the complexity of transferring these 17 genes to non-leguminous plants must be added that of transferring the genes necessary for formation of the *Rhizobium* nodules themselves.

Others[2] have argued that the energy cost to, for instance, a cereal crop of fixing the amount of nitrogen which it is currently consuming under moderately high yielding conditions (wheat producing 7t/ha typically requires 150 kg of nitrogen/ha) will be so high as to *reduce* yields by 20-30% from current levels. Nitrogen fixation among non-leguminous crops must therefore be accepted as a technology unlikely to become available before the early decades of the next century and then, at best, will produce only modest amounts of nitrogen.

Progress in identifying mechanisms for resistance to bacterial and fungal diseases has also been slow. No naturally-occurring plant resistance genes have yet been isolated and cloned.

Substantial progress seems likely in genetic engineering to improve the quality of a number of products, which on the whole do not involve complex combinations of genes. For example, some of the genes responsible for endosperm storage protein in wheat have been isolated and can be altered to test theories on how the aminoacid sequences of their protein products contribute to bread making. Other prospects include genetic engineering to introduce new fatty acids (or modify the balance of existing ones) in oilseed rape, the introduction of genes to enhance storage proteins in legumes and to improve shelf life, fruit texture and flavour in tomatoes.

Gradual increases in the efficiency of conventional plant breeding resulting from a range of new techniques are likely to be achieved in the next decade and passed on to the farmer. These include the various methods of germplasm management and disease-free propagation already being used at the IARCs for preparation and despatch of genetic material to ldc breeding programmes. More sophisticated techniques, although costly and largely experimental, include restriction fragment length polymorphism (RFLP) markers designed to identify large blocks of DNA or chromosome segments that contribute to qualitative characters such as yield. These may assist the breeder in systematically testing and assembling the building blocks of the genomes he seeks, augmenting the empirical procedures currently used. CIMMYT and IRRI have decided to investigate the implications of RFLPs for their breeding programmes and have set up research teams and laboratory facilities to pursue this, in collaboration with laboratories heavily involved in this work in the USA, Western Europe, Japan and Mexico.

Control of pests, diseases and weeds

Biotechnology offers substantial prospects for increased understanding of the characteristics of pathogens and their modes of interaction with economically useful plants. This will lead to greater efficiency in their control, either directly, or through assisting plant breeders to identify and exploit resistance mechanisms. Wide application can be expected in the future of the monoclonal antibody and c-DNA techniques recently introduced into the detection of viral, bacterial and fungal diseases. For instance, over 300 types of mycoplasma are recognised, and with conventional techniques it is difficult to separate out their individual characteristics and effects (Payne, 1988).

The incorporation of resistance mechanisms into new plant varieties is an important weapon in the search for protection against pathogens. The use of biotechnology in generating herbicide resistance, as outlined above, will be a strategy of major commercial importance, though any directly beneficial impact on ldcs is likely to be limited. The control of virus disease is unlikely to be achieved by methods other than varietal resistance. Further progress in this area requires continuing studies of host-pest genetics to identify additional genes and gene products that induce pest and disease resistance.

A different approach is offered by the development of microbial agents for pest, disease and weed control. Recent advances in genetic engineering allow improvement in some strains and the transfer of genes (eg. insect-pathogenic toxins) into other organisms which persist within the environment and have the potential to exert longer-term control on pest populations than the parent microorganism itself.

Overall, progress in these areas is constrained by lack of basic knowledge of the characteristics of pathogens, and of their interaction with crop plants. Whilst the progress being made in some areas (eg. plant breeding for resistance) will be reinforced by biotechnology, its impact on ldcs is likely to be slow and piecemeal until advances in fundamental understanding have been made.

Veterinary

Growth hormones figure largely in the first r-DNA products for the animal market, and have been developed for poultry, swine and bovine applications. These are still under trial, and although their potential market has been estimated at hundreds of millions of dollars (Hacking, 1986, p.264) they face regulatory legislation in some applications. Specific applications such as the hormone bovine somatotropin, designed to stimulate increased milk yield, is already available in the UK market.

Methods of mass production of embryos of predetermined characteristics have been established, so that embryo transfer to a surrogate mother offers strong commercial potential. The sex of embryos will be determined prior to implantation, and the time required for reproduction will be substantially shortened, so that a cow kept under intensive conditions which currently produces an average of 3.5 calves in its lifetime is expected to produce 17 under embryo transfer.

Vaccines based on r-DNA products are being developed for numerous veterinary applications, including: scours, a neonatal bacterial diarrhoea of cattle and pigs, estimated to cause the death of almost 5% of the 53 million calves produced annually in the USA, and a major problem in pig production, where conventional vaccines command a $100m market. Foot and mouth disease, although virtually eliminated from intensive agriculture through a combination of conventional vaccine, quarantine and slaughter of infected herds, is still endemic in South America, much of Asia and Africa. R-DNA vaccines are expected to gain part of the market (with several hundreds of millions of dollars because of evidence that residual virulence in present vaccines causes outbreaks of the disease. *Pseudorabies* is a herpes virus which infects the nerve cells of pigs and often remains quiescent, and therefore difficult to detect, for long periods before it reverts, causing epidemics. A commercial vaccine ('Omnivac') genetically engineered by deleting the virus's gene for allowing it to escape from nervous tissue is now available in the USA and is likely to be followed by others containing their own diagnostic kits.

The use of monoclonal antibodies in immunologic and diagnostic procedures is likely to become widely commercialised and will lead to substantial productivity gains. Although the techniques are now routine enough for eg. a university department to make its own diagnostic kits, contractual services are now becoming widely available in industrialised countries and are likely to be more cost effective for the sporadic use patterns characterising demand for this technique.

Industrial processes

Fermentation

The evolution of fermentation in food processing is reviewed by IDRC (1985), a significant stage in its 6000 year history being the discovery at the beginning of this century that microbial cells could be manipulated to change the nature and concentration of their metabolites. Strain selection and progressive improvement in the culture media have led to substantial increase in the yields of, for instance, antibiotics. The transfer and incorporation of exotic fragments of DNA allows further gains either by increasing the production of normal metabolites or by stimulating the biochemical synthesis of substances foreign to the organism.

Many of the same types of technique are used in fermentation, regardless of whether the output is intended for pharmaceutical, chemical industry, agricultural, food or energy applications. IDRC (1985) provides a list of such applications which is reproduced in Annex 3 since it provides clear evidence of the wide cross-sectoral linkages which were identified in Chapter 1 as a key characteristic of biotechnology.

Our attention here is focused, first, on a small sub-set of the range of possible applications, selected to illustrate some of the more immediate commercial applications and, second, on the difficulties of scaling up operations from the laboratory to the industrial level.

Examples from the food industry. In many parts of the world, simple fermentations produce foods directly for the consumer. Traditional beers in Africa and a range of fermented foods and sauces (including soy sauce) in South East Asia provide examples. Much of this technology hitherto has been developed and applied at the village or small-scale industry levels. Several observers (eg. Pyle, 1988) have noted that with better understanding of the microbiological process4s involved, and of nutritional implications, and with developments in process engineering, the production of fermented foods could become more efficient and expand substantially.

However, many of the products of fermentation processes are not consumed directly, but serve as inputs into other foods or food production processes. Enzymes are an important example of these, being used to improve food processing steps (eg malt enzymes), to increase process yields, and to enhance product quality (eg. clarification of fruit juices). Some enzyme processes have led to substantial efficiency gains in one type of product (eg. high-fructose corn syrup) with consequent partial displacement of others (cane sugar). The principal types of enzyme attracting substantial research include those to produce flavours and aromas (eg. esters in the price range $200-$1000/kg) and those to modify texture, appearance and nutritional quality. Potential applications of these include the processing of low-grade materials (eg. cheap saturated tallow fats) to compete with existing products such as cocoa butter, although most of these applications — including improved cocoa butter substitutes (CBS) — are far from commercial application, given the complexities and high costs of the processes involved, the high costs of transporting bulk low-grade raw materials and (especially in the case of CBS) legislation governing the permissible levels of admixture of synthetic materials. Commercial success has, however, been achieved in the production of certain enzymes and proteins through genetic engineering. Examples include rennit produced from food grade organisms and used in a wide range of applications, including cheese manufacture, and high intensity protein sweeteners such as thaumatin.

The industrial production of protein via microbial processes from non-renewable sources (eg. hydrocarbons) or from renewables such as molasses, whey and agro-byproducts has attracted much attention. Following the commercial failure of one large-scale effort (ICI's 'Pruteen') it seems clear that processes based on non-renewable sources will not be viable for US or European conditions, but may find application in oil-producing countries having limited agricultural potential. Efforts based on renewable resources (such as Rank-Hovis-McDougall's 'mycoprotein') seem more likely to be successful, but face long delays and uncertainty in meeting public health registration requirements.

Hacking (1986) provides a succinct summary of the potential that biotechnology has in fermentation processes:

'The introduction of a genetically-engineered organism into a fermentation system can improve cell productivity by increasing the expression of genes and by increasing their copy number. The system's efficiency is improved in terms of productivity, reaction ratios and substrate conversion efficiencies, thereby reducing unit costs and capital investment...In addition, genes may be transferred to a host organism which is more suited to the process and is economically more efficient. Factors here include thermotolerance, product resistance, flocculation, non-pathogenicity, tolerance of extremes of pH...lower production of wasteful by-products..'

He goes on to note that many of these goals have been achieved by conventional selection mechanisms, and that variations in eg. the ratio of raw material to processing costs from one application to another means that the scope for efficiency gains through genetic engineering varies widely. For instance, in yeast-ethanol fermentation, conversion efficiencies are already close to the theoretical maximum, and process costs are in any case small by comparison with substrate costs, so that genetic engineering could produce only limited cost savings. Similar conclusions apply to other bulk fermentation processes such as single cell protein, yeast biomass, citric acid, lactic acid, monosodium glutamate, lysine, xanthan, penicillin and riboflavin. In other (often small volume) products, however, yields, conversion efficiencies and concentrations remain low in spite of conventional research. The amino acid tryptophan could, for instance, achieve higher sales in animal feed if its production costs could be brought below $10/kg through biotechnology. Similarly, some vitamins (eg. B12) and antibiotics suffer poor conversion efficiencies and substantial cost savings could be achieved through genetic engineering, but progress has hitherto been slow because the pathways to these products are controlled by many genes with diverse chromosomal locations and regulatory systems.

Scaling up. As IDRC (1985) notes, fermentations have conventionally been carried out in batch systems in which the microorganism and its nutrient medium interact in a fermenting vessel until the metabolites generated reach maximum achievable concentrations and the process halts. The low concentrations characterising this process (eg. 10-12% ethanol for most yeast processes; under 5% for some antibiotics) have led to the development of continuous processes in which dissolved substrates are passed over microorganisms or enzymes held static in reactor columns. Some continuous fermentations involving immobilised bacteria are 200 times more efficient than batch processes, and substantial gains are expected from those involving immobilised enzymes and microorganisms, though these processes are still evolving.

The principal constraints on rapid and widespread implementation of these include (IDRC, 1985):

— the severe shortage of biological engineers (and chemical engineers' unfamiliarity with biological processes) so that the scale-up from laboratory to commercial production has to be done partly on a trial-and-error basis and even large companies have made costly errors

— the peculiarities of each type of fermentation in terms of optimum temperatures, mixing, aeration and media composition. The maintenance of aseptic conditions and even temperatures in large scale fermentations is especially problematic.

— the fact that most fermentations take place in dilute solutions, thereby requiring large quantities of uncontaminated water and, as scale-up progresses, making more costly the process of extracting the desired end-product from the microbiological biomass and other extraneous by-products.

— the increased difficulty of preventing mutation among microbial species as processes move from batch to continuous and from small to large scale.

Plant cell culture. The ability to grow plant cells on a laboratory scale has aroused much interest in varietal improvement but it has also become possible to design cultures which directly generate commercial products (mostly as secondary metabolites). However, fermentation in this application is much more expensive than for bacteria or yeasts because growth rates are slower and growth media much more expensive. Thus, only synthesis of expensive compounds such as drugs or fragrances (with a market value of $100/kg) is likely to be economic. The only product currently manufactured commercially by cell culture is shikonin, a dye and pharmaceutical used in Japan, where consumption has been some 150 kg/year or $4000 per kg. Its manufacturer (Mitsui) anticipates that the lower-priced synthetic will generate market expansion. Claims have been made, however, that work in progress will reduce the costs of culturing vanilla from its present $2000/hg to $50/hg, which would make it price-competitive with the conventionally-grown product. A recent review of techniques and their commercial potential (Hamill et al, 1987) concludes that the low and/or unstable productivity of many undifferentiated cultures may be overcome by the use of rapidly-growing 'hairy' root cultures from plants genetically transformed by *Agrobacterium rhizogenes*. The major secondary products derivable via hairy root cultures include atropine, hyoscyamine, nicotine products, steroidal alkaloids, and quinoline alkaloids. While this technique appears to have commercial potential, its application is at present restricted by limited knowledge of where and how these secondary metabolites are synthesised within plant organs.

Forestry

Genetic studies on forest species have tended to lag behind those on annual crops, and on industrial food and beverage crops. The long growth cycle of tress, the difficulty of rooting propagules from mature trees and poor juvenile-mature correlations of characteristics are long-standing impediments to genetic improvement (Burley, 1987).

Tissue culture offers strategies to reduce the time required for reproduction of trees, and to introduce targeted genetic improvements. Micropropagation is currently the technique most widely practised, but recent work in the USA has succeeded in using *Agrobacterium* as a vector for the genetic transformation of principal timber species (loblolly pine and Douglas fir),[3] which opens the broader perspectives of genetic engineering. Somaclonal and gametoclonal biotechnologies are projected to be the primary tissue culture techniques to serve tree improvement in the next 15 years (Haissing et al, 1987), being used to obtain highly targeted improvements relying on one or a few genes which can be screened in culture or in the early stages of plantlet growth. Resistance to diseases and chemical stresses (salts; herbicides) and growth rate, morphology and flowering characteristics will be the breeders' principal targets. Other possibilities include manipulation of the nitrogen-fixing mechanisms of tropical leguminous species and the use of monoclonal antibodies to screen soils for fungi likely to impair the establishment of forest plantations. In the processing of wood and straw into eg. paper, there is scope for increased efficiency in lignin breakdown through genetic engineering of certain fungi (Burley, 1987).

Institutional forms

For ***industrialised countries***, there are no immediate reasons for supposing that the pattern of heavy involvement of the private sector in biotechnology R & D will change. However, it is worth outlining the preconditions for continued investment:

First, that sufficient funding, from a combination of venture capitalists, stock markets and major commercial companies, continues to be available. Whilst the marketability of biotechnology shares appears to have been influenced by cycles of enthusiasm and disillusion on Wall Street.[4] There seems little doubt that private finance in various forms will continue to fund the bulk of biotechnology R & D, providing that: Second, patent and trade secret legislation develops sufficiently to allow companies to reap returns from their R & D investment and that commercial application of biotechnology products can meet health and safety requirements and Third, that companies generate marketable products and processes to provide investors with an adequate return. Although many biotechnology possibilities are still far from commercialisation, some products and processes in medical, veterinary and industrial fermentation are already marketed, and a small number ofted to follow in the next 5 years. The buying-in by major agrochemical companies of biotechnology expertise to address specific issues (such as breeding for herbicide resistance) increases the pressure on agricultural biotechnology towards market orientation, and substantial change in staffing structures to incorporate more marketing personnel are a move in the same direction.

The needs of *developing countries* seem likely to be met largely from publicly-funded sources, particularly the IARCs and national research services. In countries with a substantial industrial or agro-industrial base, limited amounts of private capital have been invested in biotechnology R & D, and some countries growing major agro-industrial export crops (rubber; palm oil) have been able to benefit from R & D financed by multinationals, but these remain specific instances. A wider increase in private funding for biotechnology R & D will have to wait for a healthy private sector and stronger industrial base in many ldcs.

Efforts have also been made to establish new multilateral publicly-funded institutions (ie. outside the CGIAR framework) for conducting agricultural biotechnology R & D in and for ldcs. UNIDO has taken a major initiative in this regard with the proposed International Centre for Genetic Engineering and Biotechnology (ICGEB). This was proposed in 1981 as a centre of excellence with the brief of monitoring, adapting and developing biotechnology for the particular needs of developing countries. Considerable controversy has surrounded the terms of reference of this organisation, the institutional form that it should take, its location, funding arrangements and terms and conditions for staff, with the result that by 1988 its statutes have not yet been ratified by over half of the 40 signatory countries, and no permanent accommodation has been constructed for the ICGEB in one of its two proposed locations (New Delhi). It is therefore too early to predict what impact this institution might make, but the uncertainties of the last 7 years do not augur well (Zimmerman, 1987).

In the absence of new institutional forms and of substantial private sector involvement in agricultural biotechnology R & D for ldcs, requirements must continue to be met by the two main types of agency prominent hitherto, international agriculture research centres (IARCs) and national agricultural research services (NARS).

NARS vary widely in their size and productivity. Whilst there is evidence that the productivity of some is low and declining (see a review of African NARS by Bennell and Thorpe, 1987), others have made substantial progress in biotechnology. It is reported from China, for instance, that researchers have achieved reproduction of rice through anther culture (Joffe and Greeley, 1987). Substantial (publicly-funded) progress has been made in India with germplasm conservation through tissue culture techniques at the National Facility for Plant Tissue Culture Repository under the National Bureau for Plant Genetic Resources. This is responsible for research on *in vitro* conservation, including cryopreservation of seeds, anthers, pollen grains, calli and organs.

Molecular biology and genetic engineering research in India are undertaken at the Centre for Cellular and Molecular Biology, the Indian Institute of Chemical Biology and the National Institute of Immunology. Tissue culture is a primary area of research at the Council of Scientific and Industrial Research, and the National Botanical Research Institutes. Training in tissue culture is offered at several of the universities, and in biochemistry and microbiology. The Indian Institute of Technology offers formal courses in biochemical engineering. The National Biotechnology Board aims to coordinate research and training activities around the priority areas of genetic engineering, photosynthesis and tissue culture (Belay, 1988, quoting Swaminathan).

In spite of these efforts, India appears to suffer difficulties likely to be common among ldcs. Dembo et al (1987), for instance, report a case of conflicting interests in which an attempt to produce rare medicinal plants by cell culture appeared to undermine the livelihoods of those engaged in gathering and processing the wild material. Furthermore, the laboratory techniques in many branches of biotechnology, whilst sophisticated enough to require PhD-level skills, are tedious and repetitive, and so are unlikely to be compatible with the aspirations of returning highly-qualified ldc nationals. This may in part account for the high numbers of Indian researchers engaged in biotechnology research elsewhere — currently estimated at 400 in the USA and over 100 in Europe (Belay, 1988).

Whilst mutually reinforcing exchange of material and techniques is possible between certain NARS and the IARCs, in other cases the interaction is likely to be limited to the supply by IARCs of improved material for conventional breeding techniques. Substantial financial support is necessary if NARS are to achieve the critical minimum mass of research effort to evaluate this material effectively. A summary of the principal biotechnology activities at the IARCs is presented in Annex 2.

The complementarity between conventional and biotechnology-based methods of plant breeding has been stressed by a number of observers (Cohen et al, 1988; Vasil, 1988). Broadly, tissue culture techniques facilitate the arrangement of germplasm collections and the rapid supply of disease-free material of known characteristics. Gene insertion techniques broaden the range of genetic variation from which breeders can make crosses, being able (in as yet a limited number of cases) to insert genes to produce known plant traits. Yet, the search for, or creation of, genetic diversity is the less arduous component of plant breeding. It is the selection of useful characters from the available diversity which is more time consuming, and still has to be performed with material containing genetically engineered components, since knowledge of the types of trait produced by particular genes is far from complete, and, on the whole, cross-bred plants have to be grown out to maturity in successive cycles of breeding in desirable characteristics and breeding out undesirable ones by conventional methods.

The foregoing discussion has emphasised the fact that the earliest commercial applications of the more fundamental advances in agricultural biotechnology (through 'genetic engineering') are likely to be some 5-10 years from now. With certain major food crops (particularly the cereals), and with the types of plant modification likely to give greatest productivity increase, the lead-time is likely to be longer. On the other hand, certain technologies in the broad field of tissue culture are already being used as adjunct to conventional plant breeding, and the productivity gains they offer are likely to come on-stream within 5 years.

Impact on ldcs

Emerging from the literature is an awareness that certain fallacies could easily distort ldc policy towards biotechnology. Dembo et al (1987, p 17), for instance, argue that these include:

— the 'innocent bystander/wait-and-see' fallacy; in varying degrees, biotechnology will affect *all* ldcs.

— fallacies relating to *time*, which imply that there is no need to act now, since the effect of many biotechnologies will not be felt for at least, one possibly two, further decades. But pressure for change must be exerted now before certain policy options, especially relating to the withdrawal of materials and research results from the public domain, are foreclosed.

— the TCDC (technical cooperation among developing countries) fallacy which suggests that the more advanced ldcs will be ready to assist others in biotechnology. Given the highly competitive environment in which it is being developed, this is unlikely to materialise.

They also argue that the IARCs are unlikely to be as successful in promoting biotechnology as in earlier Green Revolution technologies unless ldcs promote strategies to strengthen these institutions. They seem to imply here a degree of political support. Whilst such support is politically important, in practical terms it is likely to have little impact, given that the bulk of finance from the IARCs comes from OECD countries. What is, however, important is that ldcs, through their national agricultural research systems, should be able to exert some 'demand pull' on the IARCs, to enhance the relevance of the technologies they develop to ldc conditions.

Several types of impact on ldcs are conceivable:

i. on crop productivity in ldcs through direct use of the technologies
ii. on trade patterns between ldcs and industrialised countries resulting from differences in the timing and magnitude of productivity increases both in agriculture and in agro-industry between the two groups of countries
iii. on nutrition with ldcs through improvements in the quality of agricultural and agro-industrial products.

However, given the uncertainties surrounding the size, location and timing of implementation of individual technologies, it is extremely difficult to predict in any detail the productivity gains that may be achieved, and their impact on trade. The following are relevant considerations:

1. The macro agro-economic environment will influence the potential rate of uptake of biotechnologies. The depressed condition of US agriculture through problems of indebtedness and over-production, and of European agriculture through surpluses are seen by the biotechnology industry as an unfavourable environment for most of the technological innovations likely to come onto the market.[5]

2. The distortions in trade between North and ldcs resulting from protectionism and from the generation and disposal of surpluses in major food commodities are substantial, possibly amounting to a net current cost to ldcs of $700m/yr, (Matthews, 1985) and unless resolved, will greatly outweigh the price and trade effects of any productivity gains resulting from biotechnology in the next 10-15 years.

3. Where productivity gains in the North do occur in traded commodities as a result of improvements in biotechnology, the impact of this on ldcs will depend on whether they are importers or exporters of the commodity in question. Exporters stand to lose through reduction in world market prices, but importers stand to gain. Econometric analyses of impacts in price fluctuations in major commodities on ldcs have been published (eg. Matthews (1985)) and

give an impression of the order of magnitude involved, even though they principally concern price changes induced by modifications to the Common Agricultural Policy, not through biotechnology. Whilst expenditure on biotechnology R & D is greater in the North than in and for ldcs, many of the productivity gains being sought in the North involve the more complex genetic engineering aspects of biotechnology, which are unlikely to become commercially available for major traded commodities for at least 10 years.

4. The less ambitious aspects of biotechnology (ie. tissue culture techniques) are being researched in the North, but are a particular focus of research in and for ldcs, as a complement to conventional breeding techniques. Generalisations are difficult to support, but it seems likely that their potential in this mode is greater in ldcs than in the North. Commercial crops in the North have been the focus of much greater attention from breeders over several decades than have important ldc food crops. Present yields in the North are therefore likely to be close to their potential, whereas in ldcs, tissue culture techniques have an important role in assisting breeders to exploit the substantial remaining gap between actual and potential yields. A further strand to this argument is that most of the North's crops are grown under highly homogeneous conditions, whereas a large proportion — perhaps a majority (Chambers and Jiggins, 1987) — of ldc producers operate under highly variable agro-ecological conditions. The acceleration of conventional breeding processes facilitated by tissue culture will allow breeders to give more attention than hitherto to producing plants with the specific characteristics best suited to these varied conditions. An important consideration here is that techniques are only now being developed to elicit from farmers in these difficult conditions the priorities that they would attach to agricultural research designed to relieve the constraints they face, and to match these with the opportunities offered by the body of accumulated scientific knowledge (Farrington & Martin, 1988). NARs will require considerable strengthening if they are to fulfil this role successfully and, at the same time, to adapt conventional and biotechnologies to meet these requirements.

5. Attempts have been made for decades to synthesise substitutes within the 'North' for products generally imported from ldcs. Synthetic rubber and fibres have had major commercial impact as have synthetic flavourings, to a lesser degree, but the import

substitutes derived through biotechnology have so far been modest in scope.

Plant cell culture at the agro-industrial level has so far yielded only shikonin as an important substitute, and the scope for cost reduction in this process offered by improvements currently under investigation is limited, so that it is unlikely to produce anything other than high value ($100/ha) and low-volume substitutes in the next decade.

Fermentation processes offer variable scope for improvement through genetic engineering. Conversion efficiencies are already high in bulk fermentations, and processing represents only a small proportion of total costs, so that the scope for improvement is limited. This applies to yeast biomass, citric acid, lactic acid, monosodium glutamate, lysine, xanthan, penicillin and riboflavin. Only in the higher value amino acids, vitamins and antibiotics (ie. over $10/kg) are substitutions likely to be made in the next 5-10 years.

6. An important development in the fermentation industry is the capacity to use an increasing range of agricultural products (or by products) as raw material for extraction of specific chemicals, switching from one to another as market conditions change. Whilst it is too early to predict the impact of these developments on ldcs, it seems clear that some markets (for vegetable oils) will become more volatile perhaps increasing the disadvantage faced by producing countries which do not have easy and rapid routes to world markets.

7. Fermentation is a widely-practised traditional means of food preparation in ldcs, but with some exceptions in E. Asia, is found at the village, not agro-industrial, level. The prospects for uptake of new fermentation processes by ldcs appears limited by lack of a tradition of agro-industrial fermentation and, in smaller countries, limited markets for their output. Economies of scale dictate high capital cost and high output for many fermentation processes (in the order of $250,000-$500,000), though progress has been made recently in scaling down the equipment used in some processes (Aronsson, 1987; van Brunt, 1987)

8. Some nutritional impact in ldcs from genetic engineering is to be anticipated in the next decade. Synthesising a DNA sequence that will either code for a synthetic protein or carry out a regulatory protein-producing role is relatively easy and could add major nutritional improvements, raising protein levels in, for instance, root crops during storage. It is under investigation in the IARCs.

In general terms, we must conclude that it is too early to predict in any detail what the impact of biotechnology will be on ldcs. On present evidence, biotechnology-induced changes in trade patterns are likely to be small in the next decade, particularly when compared with current distortions resulting from protectionist policies in the North. Some improvement in yield and nutritional value of ldc food crops can be expected over that period, but much will depend on the capacity of ldc national research institutes to incorporate the genetic material exhibiting these characteristics into breeding and multiplication programmes.

Ultimately, no ldc will be able to isolate itself from the trade effects resulting from the application of biotechnology to the food industry. Nor is there room for complacency in view of the likely lag in commercial application of many of the technologies now in prospect: a fine balance in policy and legislation will soon have to be achieved if commercial companies are to be prevented from foreclosing the options of most interest to ldcs (chiefly relating to open access to the results of research) without such disincentives as to reduce research and commercialisation to an unacceptably slow pace. Although recognised as essential in many ldcs, the strengthening of national agricultural R & D capacity has not been given adequate priority in others, and, at practically all levels — whether at the International Research Centres, the ldc national service or in the creation of new regional centres — it is difficult to envisage substantial progress without high levels of donor assistance in the development of skills, infrastructure and techniques.

Notes

1. *Bio/Technology*, Vol 5, February 1987, pp 128-133
2. Reeves, personal comment
3. *Bio/Technology*, Vol 5, February 1987
4. *Bio/Technology*, Vol 5, May 1987, p 429; *Ibid*, Vol 6, March 1988, p 243
5. *Bio/Technology*, Vol 5, February 1987, pp 128-133

Bibliography

Aaronson, G. 'Plant equipment: scaling down for downstream scale-up'. Bio/Technology Vol 5, April 1977, pp 394-395.

Arnold, M H. 'Biotechnology and priorities for agricultural research'. Paper presented to the CGIAR. mimeo, undated.

Belay, A. 'Is new biotechnology a threat or opportunity to Third World agriculture?'. MSc Research Policy, Sweden: University of Lund, 1988.

Bennell, P & Thorpe, P. 'Crop science research in sub-Saharan Africa: a bibliometric overview'. *Agricultural Administration and Extension*, 25 (2) 1987, pp 99-123.

Binns, M M; Tomley, F M; Campbell, J & Boursnell, M E G. 'Comparison of a conserved region in fowlpox virus and vaccinia virus genomes and the translocation of the fowlpox virus thymidine kinase gene'. *J gen Virol*, 69, 1988, pp 1275-1283.

Blajan, L. 'World production and utilisation of products of animal origin'. *Rev Sci Tech Off Int Epiz*, 6, 1987, pp 849-883.

Bodet, P. 'Animal energy'. *World Animal Review*, 63, 1987, pp 2-6.

Bonga, J H. 'Vegetative propagation in relation to juvenility, maturity and rejuvenation', in: J M Bonga, D J Durzan eds. *Tissue Culture in Forestry*, The Hague: Martinus Nijhoff/Dr W Junk, pp 387-412.

Boyle, D B & Coupar, B E H. 'Construction of recombinant fowlpox viruses as vectors for poultry vaccines'. *Virus Research*, 10, 1988, pp 343-356.

Brinster, R L; Allen, J M; Boehringer, R R; Gelinas, R E & Palmiter, R D. 'Introns increase transcriptional efficiency in transgenic mice'. *Proceedings of the National Academy of Science*, US, 85, 1988, pp 836-840.

Bryan, J E. 'Implementation of rapid multiplication and tissue culture methods in third world countries'. *American Potato Journal*, 65, 1988, pp 199-207.

van Brunt, J. 'A closer look at fermentors and bioreactors'. Bio/Technology Vol 5, November 1977, pp 1134-1138.

Bull, A T; Holt, G & Lilly, M D. *Biotechnology — international trends and perspectives*. Paris: OECD, 1982.

Burley, J. 'Applications of biotechnology in forestry and rural development'. *Commonwealth Forestry Review*, 66(4), 1987, pp 3357-367.

Chambers, R, & Jiggins, J. 'Agricultural research for resource-poor farmers: a parsimonious paradigm'. Discussion Paper 220, IDS, University of Sussex, 1986.

Chandler, S F & Thorpe, T A. 'Variation from plant tissue cultures: biotechnological application to improving salinity tolerance'. *Biotech. Adv.* 4, 1986, pp 117-135.

Clarke, A J; Simons, P; Wilmat, I & Lathe, R. 'Pharmaceuticals from transgenetic mice'. *Trends in Biotechnology,* 5, 1987, pp 20-24.

Chilton, M D. 'Plant Genetic Engineering: Progress and Promise'. *J. Agric. Food Chem.* 36, 1988, pp 3-5.

CIP. *Annual Report 1986-87.* Lima, Peru, 1987.

Cocking, E C. 'Somatic Hybridization: implications for agriculture'. In: *Biotechnology in Plant Science.* (M Zaitlin et al. eds), New York: Academic Press, 1985, pp 101-113.

Cohen, J I, Plucknett, D L, Smith, N J H & Jones, K-A. 'Models for integrating Biotech into crop improvement programs'. Bio/Technology Vol 6, April 1988, pp 387-392.

Dembo, D; Dias, C & Morehouse, W. 'Biotechnology and the Third World: caveat emptor'. *Development* 4, 1987, pp 11-18.

Diallo, A; Barrett, T; Barbron, M; Subbarao, S M & Taylor, W P. 'Differentiation of rinderpest and peste des petits ruminants viruses using specific cDNA clones'. *J Virol Meths.* In Press, 1988.

Dimarchi, R; Brooke, G; Gale, C; Cracknell, V; Doel T & Mowat, N. 'Protection of cattle against foot-and-mouth disease by a synthetic peptide'. *Science* 232, 1986, pp 639-641.

Farrington, J & Martin, A M. 'Farmer Participation in Agricultural Research — a review of concepts and practices', Occasional Paper No 9. London: ODI, 1988.

Ferris, N P; Powell H & Donaldson, A I. 'Use of precoated immunoplates and freeze-dried reagents for the diagnosis of foot-and-mouth disease and swine vesicular disease by enzyme-linked immunosorbent assay (ELISA)'. *J Virol Meths*, 19, 1988, pp 197-206.

Fox, J L. 'The US regulatory patchwork'. Bio/Technology Vol 5, December 1987, pp 1273-1277.

Freeman, B M & Bumstead, N. 'Breeding for disease resistance — the prospective role of genetic manipulation'. *Avian Pathology* 16, 1987, pp 353-365.

Fujii, J A A, Slade, D T, Redenbaugh, K & Walker, K A. 'Artificial seeds for plant propagation'. *Tibtech* 5, 1987, pp 335-339.

Gibbs, J N. 'Exporting biotechnology products: a look at the issues'. Bio/Technology Vol 5, January 1987, pp 46-51.

Glosser, M W. 'Regulation and application of biotechnology products for use in veterinary medicine'. *Rev sci Tech Off* Int Epiz 7, 1988, pp 223-237.

Grimsley N et al. *Nature*. London, 325, 1987, p 177.

Hacking, A J. *Economic aspects of biotechnology*. Cambridge, 1986.

Hamill, J D, Parr, A J, Rhodes, M C J, Rubins, R J, & Walton, N J. 'New Routes to plant secondary products'. Review article. Bio/Technology Vol 5, 1985, pp 800-804.

Hobbelink, H. *Biotechnology and Third World agriculture: new hope or false promise?* Brussels: International Coalition for Development Action, 1987.

Horsch, R B, Fry, J E. Hoffman, N L, Eichholtz, D, Rogers, S G & Fraley, R T. 'A simple and general method for transferring genes into plants'. *Science*, 277, 1985, pp 1229-1231.

Howard, B H & Marks, T R. 'The in vitro-In vivo interface'. In: *Advances in the Chemical Manipulation of Plant Tissue* (M B Jackson, S H Mantell, Jennet Blake eds). British Plant Growth Regulator Group, Monograph No. 16, 1987, pp 101-111.

IDRC. 'Biotechnology: opportunities and constraints'. Manuscript Report IDRC-M110e Ottawa, 1985.

Joffe, S, & Greeley, M. 'New plant biotechnologies and rural poverty in the third world'. Unpublished report commissioned by Appropriate Technology International. Institute of Development Studies, University of Sussex, 1987.

Jonard, R. 'Micrografting and its applications to tree improvement', in: Bajaj, I.Y.P.S. ed, *Biotechnology in Agriculture and Forestry Vol 1: Trees*, Berlin, Heidelberg: Springer-Verlag. 1986, pp 31-48.

Juma, C. *The Gene Hunters*. London: Zed Books, 1988.

Kirsop, B E. 'Culture collections: repositories for microbial germplasm'. *Nature and Resources* 23, 2, April-June 1987, pp 2-9.

Klausner, A & Fox, J. 'Some bird's eye views of agbiotech '88'. Bio/Technology Vol 6, March 1988, pp 243-244.

McCoy, T J. 'Characterization of alfalfa (Medicago sativa L) plants regenerated from selected NaCl-tolerant cell lines'. Plant Cell Reports 6, 1987, pp 417-422.

Mantell, S H & Hugo, S A. 'International germplasm transfer using micropropagules', in: P Kapoor & S H Mantell eds. *Biological Diversity and Genetic Resources Techniques and Methods: Mass Propagation using Tissue Culture and Vegetative Methods*, CSC Technical Publication Series No. 205, 1986, pp 88-98.

Mantell, S H, Matthews, J A & McKee, R. *Principles of Plant Biotechnology*. Oxford: Blackwell Scientific, 1985.

Marx, J L. 'Rice plants regenerated from protoplasts'. *Science* 235, 1987, pp 31-32.

Matthews, A. *The Common Agricultural Policy and the Less Developed Countries*. Dublin: Trocaire, 1985.

Moses, P B; Tavares, J E & Hess, C E. 'Funding agricultural biotechnology research'. Bio/Technology Vol 6, February 1988, pp 144-148.

Moss, B; Fuerst, T R; Flexner, C & Hugin A. 'Roles of vaccinia virus in the development of new vaccines'. *Vaccine* 6, 1988, pp 161-163.

Murray, P K. 'Prospects for molecular vaccines in veterinary parasitology'. *Veterinary Parasitology* 25, 1987, pp 121-133.

Nabors, M W; Gibbs, S E; Bernstein, C S & Meis, M E. 'NaCl-tolerant tobacco plants from cultured cells'. *Z Pflanzenphysiol.*, 97, 1980, pp 13-17.

Navarro, L & Juarez, J. *Tissue culture techniques used in Spain to recover virus-free Citrus plants*. Acta Hortic. 78, 1977, pp 425-435.

Newmark, P. 'Discord and harmony in Europe'. BIO/TECHNOLOGY Vol 5, December 1987 pp 1281-1283.

Overseas Development Institute. 'Agricultural Biotechnology and the Third World'. Briefing Paper. September 1988. London: ODI.

Payne, C C. 'Biotechnology in relation to pests diseases and weeds'. Paper presented at the Overseas Development Administration Natural Resources Advisers' Conference, University of Warwick, 10-13 July 1988.

Payne, R W J. 'The emergence of trade secret protection in biotechnology'. Bio/Technology Vol 6, February 1988, pp 130-131.

Pyle, D L. 'The impact of biotechnology on food and nutrition'. Paper presented at the Overseas Development Administration Natural Resources Advisers' Conference, University of Warwick, 10-13 July 1988.

Rhodes, C A et al. 'Genetically transformed maize plants from protoplasts'. *Science* 240, 1988, p 204.

Roca, W M, Rodriguez, J, Beltrau, J, Roca, J & Mafla, C. 'Tissue culture for the conservation and international exchange of germplasm'. In: *Plant Tissue Culture*, (A Fujiwara ed). Tokyo: Maruzen, 1982, pp 771-2.

Schaefer, W, Gorz, A & Kahl, G. 'T-NDA expression and integration in a monocotyledonous crop plant'. *Nature* 327, 1987, pp 529-532.

Scowcroft, W R. 'Somaclonal variation: the myth of clonal uniformity'. In: *Genetic Flux in Plants Springer* (Hohn, B & Dennis, E S eds). Vienna/New York, 1985, pp 217-245.

Schillito, R D; Saul, M W; Paszkowski, J; Mueller, M & Potrykus, I. Bio/Technology 3, 1985, p 1099.

Smith, C. 'Genetic improvement of livestock, using nucleus breeding units'. *World Animal Review* 65, 1988, pp 2-10.

Tartaglia, J & Paoletti, E. 'Recombinant vaccinia virus vaccines'. *Trends in Biotechnology* 6, 1988, pp 43-46.

Vasil, I K. 'Progress in the regeneration and genetic manipulation of cereal crops'. Review article. Bio/Technology Vol 6, April 1988, pp 397-401.

United States Department of Agriculture. *1986 Yearbook of Agriculture.* 'Research for Tomorrow', USDA Publication, 336 pp.

Wagner, H-G, R. 'Present status of embryo transfer in cattle'. *World Animal Review* 64, 1987, pp 2-11.

Williams, M. 'Technology, surplus food and changing farm and trade policies'. *Development* 4, 1987, pp 5-7.

Yano, S; Ogawa, M & Yamada, Y. 'Plant formation from selected rice cells resistant to salt', in: A Fujiwara ed. *Plant Tissue Culture*, Tokyo: Maruzen, 1982, pp 495-496.

Zimmerman, B K. 'UNIDO's attempt to create a 'centre of excellence' (Part I) and 'The ICGEB: can UNIDO create a centre of excellence?' (Part II) *Biofutur*, No 61, October 1987, pp 43-48 and No 63, December 1987, pp 55-62.

Annex 1
Biotechnology Glossary[1]

This Annex is an edited version of Appendix XII of IDRC (1985). Permission to reproduce is gratefully acknowledged.

Amino Acids: The chemical units of which all proteins are composed. Of the many in existence, only 20 are common to the proteins of living organisms. Of these, 11 are 'essential' to humans in that they must exist in the diet and cannot be synthesised *in vivo*.

Anabolism: Chemical changes in living organisms by which chemical energy is stored in chemically complex substances.

Anther: The fertile segment which surmounts the stamen (the male organ of a plant) and carries the pollen sacs. When the anthers are ripe they open to release the pollen.

Anther Culture: A culture of plant cells derived from excised anthers.

Antibiotics: Metabolites produced by microorganisms (and probably by higher organisms) which inhibit the growth of rival organisms. The term is particularly applied to isolated, purified and chemically modified substances isolated from cultured microorganisms and employed for therapeutic purposes. Though most widely used to describe anti-bacterials, scientifically 'antibiotics' include anti-fungal, anti-viral and anti-parasitic compounds of biological origin. Close to 100 different antibiotics for pharmaceutical applications are manufactured by industrial fermentation.

Antibody: A protein manufactured by an organism's immune system to counteract the effect of an antigen. It confers immunity against subsequent infection, in some instances permanently, in others for a limited period.

Antigen: A substance foreign to a host living organism that stimulates the immune system to generate a corresponding antibody which reacts with and renders the antigen ineffective.

Apical Meristem: The tip of a growing plant root or shoot composed of cells from which subsequent growth develops.

Asexual Reproduction: Reproduction of a plant or animal without fusion of male and female gametes. It includes vegetative propagation, cell and tissue culture.

ATP: (adenosine triphosphate): A substance present in all living organisms. Its conversion to the di- or mono-phosphate liberates energy used for many organic functions including muscular contraction, respiration and 'nitrogen fixation'.

Bacterium: A major class of unicellular organisms, the smallest living things able to reproduce themselves, achieved mainly by each cell dividing into two. The genetic code is carried in a tangled coil of DNA known as the bacterial chromosome.

Biomass: The total mass of material resulting from the growth of an organism, or organisms, plant, or animal.

Biopolymers: Long chain molecules synthesized by living organisms. Proteins are polymers of amino acid monomers, cellulose and starch are built from sugar monomers.

Callus Culture: A mass of undifferentiated cells originating from any type of explant. In a callus, usually developed on nutrient agar, the cells are generated in an unorganized clump, analogous to a pile of loose bricks. In a plant (as are bricks in an architectured building) the cells are differentiated and organized systematically to form shoots, roots and other organs.

Catabolism: Metabolic processes that liberate energy, eg. the breakdown of complex organic molecules by living organisms to liberate energy. (See Metabolism and Anabolism.)

Cell Culture: A group or colony of cells propagated from a single cell in a specifically formulated nutrient medium.

Cell Fusion: The fusing together of two or more cells to become a single cell.

Chimera: An organism or piece of DNA constructed from at least two different species.

Chromosome: A thread-like body found in cell nuclei, comprised of genes arranged in linear order. In higher organisms chromosomes consists of DNA in association with protein. In bacteria they exist as 'naked' DNA. While genes are the units of heredity, chromosomes are the units of transmission from one generation to the next. During cell division chromosomes may break, rejoin or cross over giving rise to new genetic combinations.

Clone: A collection of genetically identical cells or organisms derived asexually from a common ancestor. All members of a clone are identical in genetic composition.

Coding Sequence: The region of a gene that is *expressed* i.e. translated into protein. Also called an **'exon'**.

Cotyledon: The embryo leaf in a flowering plant. The Angiosperms (flowering plants in which seeds develop and mature inside a closed ovary) are divided into Monocotyledonae (monocots) with one seed leaf and Dictoledonae (dicots) with two.

Cytoplasm: The semi-fluid content of each cell which surrounds the nucleus (the nucleoplasm).

Diploid: see Meiosis.

DNA (De-oxyribonucleic acid): The macromolecular polymer which carries the genetic hereditary message and controls all cellular functions in most forms of life. The twin strands, in the form of a helix, are composed of successive units of the sugar de-oxyribose, phosphate and the bases adenine, cytosine, guanine and thymine, through which the twin strands are cross-linked: adenine to thymine and cytosine to guanine.

Embryo: An organism in its earliest stage of development usually surrounded by protective tissue. The young sporophyte which results from the union of male and female cells in a seed plant (also called a seed germ). An immature organism before it emerges from the egg or the uterus of the mother.

Embryogenic Callus: Callus cultures that under suitable conditions are capable of producing embryos (ie. young plants). Conversely, callus cultures that have lost this capability are termed 'non-embryogenic'.

Embryo Rescue (also termed embryo capture): When cross-pollination occurs between genetically widely different plants, the resulting embryo may be aborted because of parental mutual incompatibility. Such embryos may be excised and grown on a congenial medium such as nutrient agar. This process is called embryo rescue.

Endonuclease: Nucleases are enzymes that break down nucleic acids into strands of DNA. 'Endo' or inside nucleases act at points along the strand and thus break DNA into short pieces. Endonucleases recognize a particular base sequence in DNA and cut the DNA. Some endonucleases cut the DNA at a specific point; others appear to split the DNA sequences at random. The specific endonucleases are the tools of the genetic engineer who seeks to excise strands of DNA coded for a desirable genetic character. Endonucleases are classed as *Restriction enzymes* since they are employed, for example, by bacteria to restrict infection by viruses (bacteriophages). The bacterial restriction enzyme attacks the DNA of the infecting organisms.

Enzymes: Specific proteins which act as biological catalysts to stimulate essential biochemical reactions in all living organisms. Enzymes may be biologically synthesised, extracted and employed to catalyze laboratory or industrial biochemical reactions.

Exonuclease (see also 'endonuclease'): Nucleases that attack the nucleic acid strand starting from the ends and thus gradually shorten it are termed 'exo' or outside nucleases.

Explant: A small piece of a living tissue taken for the purpose of establishing an *in vitro* culture. Cell cultures are often identified by the source of the initial explant: e.g. meristem tip cultures, anther cultures.

Expression: see **Gene Expression**

Fermentation: The process of growing a selected organism, usually a bacterium, mould or yeast, on a substrate so as to bring about a desired change or to generate products of the cells' metabolism (eg. ethanol and carbon dioxide from yeast fermentation). The term is also used to describe biochemical conversions brought about by isolated enzymes.

Gamete: A mature sex cell or germ cell (in plants the ova and pollen grains), usually haploid in chromosome number, and capable or uniting with another gamete of the opposite sex to form a new plant or animal.

Gene: The linear units of heredity transmitted from generation to generation during sexual or asexual reproduction. In modern molecular biology each gene is a segment of nucleic acid carried in the DNA encoded for a specific protein. More generally, the term 'gene' may be used in relation to the transmission and inheritance of particular identifiable traits.

Gene expression: Evidence or manifestation of a genetically controlled characteristic. All of the chromosomal genes in an organism are by no means active at all times. In a plant nucleus as little as 5% of the DNA may be producing protein at any one time. Each organ's system has a unique set of genes, genes that are expressed only in that organ. For example, the petals and leaves of higher plants contain about 7,000 genes which express themselves in a highly specific and regulated manner. Thus all genes may be 'active' or 'silent' and the manner in which they are switched on and how the 'on-off' switches are regulated is yet to be determined by molecular biologists. The process of introducing foreign genes and switching them on in higher organisms is more complex and difficult than in microorganisms.

Gene Mapping: Determining the relative locations of different genes on a given chromosome.

Genetic Code: DNA strands are made up of 'triplet codons' (units of three nucleotides) each of which selects for a specific amino acid for inclusion in a protein strand that is being synthesized. The relation between the triplet codons on the nucleic acids and the amino acids in the proteins synthesised is known as the genetic code.

Genome: The entire hereditary message of an organism. The total genetic composition of the chromosomes in the nucleus of a gamete.

Genotype: A group or class of organisms that share a common specific genetic constitution.

Gene Vector: See **vector**.

Germplasm: Often synonymous with 'genetic material' it is the name given to seed or other material from which plants are propagated. An early theory of inheritance advanced the notion that hereditary characters were contained in an immutable 'plasm' transmitted unchanged from parent to offspring (literally (Greek): a plasm is a mould or matrix in which materials may be cast or formed: a 'plasma' is the result). A germplasm bank is an organized collection of seed or other genetic material (each genotype entered being called an accession) from which new cultivars may be generated. In a zoological context germplasm banks would include collections of preserved sperm or ova and in some cases the animals from which they are derived.

Haploid: A cell containing half the number of chromosomes present in the somatic cells. Haploidy is a characteristic of sex and germ cells (see also Meiosis).

Heterozygous: An organism that for any given character possesses different genes inherited from the male and female parents.

Homozygous: Organisms which have inherited a given genetic factor from both parents and which therefore produce gametes that are genetically stable.

Hormone: A chemical messenger secreted by the endocrine or ductless glands carried in the blood stream from the gland to a target organ. The hormone induces a specific response from that organ. For example, adrenaline stimulates the heart; auxins and cytokinins in plants stimulate cell proliferation and growth.

Hybrid: A cross between organisms that have different genomes. Hybrids are most commonly formed by sexual cross-fertilization between compatible organisms, but techniques for the production of hybrids from widely differing plants are being developed by cell fusion and tissue culture.

Hybridoma: A hybrid cell resulting from the fusion of a tumor (cancer) cell and a normal cell such as a lymphocyte from the spleen. The fused cells can be cloned and, being derived from a simple spleen cell, will secrete a pure antibody. (see **monoclonal antibodies**).

Induction: The process that causes a virus earlier inserted into the host cell's DNA to break free and to multiply.

Intergeneric Hybrids: Hybrids derived from crossing two species, each of a different genus. A viable example is triticale, a hybrid of wheat of the genus *Triticum* and rye of the genus *Secale*. The more distant the relation between the two genera, the greater the difficulty of intergeneric hybridization. Many intergeneric hybrids display a mulish infertility, unable to reproduce themselves. More than a century elapsed from the time triticale was first reported and the generation of a fertile triticale hybrid.

***In Vitro*:** Literally 'in glass'. Experimental reproduction of biological processes in isolation from a living organism.

***In Vivo*:** Biological processes within a living organism.

Meiosis: The process of division of sexual cells in which the number of chromosomes in each nucleus is reduced to half the normal number found in normal somatic cells. When two sexual cells fuse, each contributes its half of the chromosomes. The resulting embryo contains the full chromosome complement. Cells with half the chromosomes are called haploids: those with the normal chromosomal complement, diploids.

Meristem: The tip of a growing plant shoot or root. A localized group of rapidly reproducing cells at a location of active growth.

Meristem Culture: A cell culture developed from a small portion of the meristem (growing tip) tissue of a plant. Either a stem shoot or root meristem can be used.

Messenger RNA (m-RNA): m-RNA carries the genetic code for a protein from the DNA to the ribosomes where the code is read and the protein manufactured.

Metabolism: The total sum of the chemical and physical changes constantly taking place in living matter.

Metabolite: A product of metabolism.

Mitochondria: Filamentous protoplasmic bodies present in cell cytoplasm sometimes called powerhouses of the cell. They carry enzymes which catalyse the biochemical processes of cell respiration and the anabolic conversion of simple substances into compounds which store chemical energy.

Mitosis: The process during somatic cell (i.e. non-sexual cell) division by which the nucleus of each daughter cell contains a set of chromosomes equal in number to the parent cell.

Modification Enzyme: The counterpart of a restriction enzyme. It chemically modifies some of the bases so that the restriction enzyme can no longer cut the DNA.

Monoclonal Antibody: An extremely pure antibody derived from a single clone of an antibody-producing cell. Invading pathogens, viral or bacterial, carry a large number of different antigens each capable of stimulating the host's immune system to generate a corresponding antibody. A single spleen cell exposed to a specific antigen can be fused with a myeloma (cancer) cell. The resultant fused cell, called a hybridoma, continually produces an antibody specifically directed against the antigen. It will therefore seek out and identify the specific antigen. Hybridomas can be cloned and cultured to produce quantities of the pure 'monoclonal antibody'. Because of its specificity each monoclonal antibody may be used for diagnostic or therapeutic purposes. Most research and development has employed mouse antibodies grown in the peritoneal cavity of immunosuppressed mice. Recently human-human monoclonals have been studied to produce therapeutically useful antibodies for example against tetanus antigens.

Mutagen: A chemical or physical agent which brings about mutation.

Mutant: Organisms, one or more of whose properties differ significantly from the parent organism from which it was derived. An inheritable change in an organism by alteration of the genetic material; a change in the sequence or chemistry of the purine or pyrimidine bases contained in DNA molecules. Mutation can be induced by high energy irradiation or by certain chemical substances.

Mutation: A change in an organism's identifiable characteristics brought about by an alteration in one or more genes.

Mycorrhiza: A symbiotic association between a fungus and a higher plant which increases the plant's capacity to absorb nutrients from the soil.

Nitrogen Fixation: Conversion of gaseous atmospheric nitrogen to an oxidized or reduced form (ie. NO>2 or NH>3) that can be utilized by plants.

Nuclease: An enzyme which catalyzes the hydrolysis of nucleic acids.

Nucleic Acid: A chain of sugars and phosphates, with a base attached to each sugar. The sequence of these bases make up the genetic code.

Nutrient Medium: A liquid broth or semi-solid jelly containing nutrients which stimulate and sustain the culture and proliferation of bacteria, higher plant cells, or animal tissue.

Pathogen: Any disease-producing organism.

Ploidy: The number of sets of chromosomes present in an organism or cell.

Pollen: The dusty, sticky material contained in pollen sacs at the end of the anthers in a plant. When ripe, the sacs split open to release the pollen. Each ripe pollen grain contains two male nuclei equivalent to male gametes.

Pollen culture: A culture of plant cells derived from pollen in a synthetic medium (Similar to anther culture). The culture of pollen or anthers on a synthetic medium generates progeny with a single set of chromosomes. A useful means of producing homozygous plants. The single set of chromosomes being doubled by colchicine.

Pollination: The transfer of pollen from the male anther to the female stigma of a flower. Pollen carried between anther and stigma of the same flower is called self-pollination. Pollen carried from the flower of one plant to another of the same species is called cross-pollination. The pollen passes through the pollen tube and the ovary into the ovule where it fertilizes the egg cell. Fusion of the male and female gametes develops in the ovule into a seed consisting of the embryo plant and its nutrients in a protective coating.

Polymer: A macro-molecular substance made up of many repeating smaller units (monomers) bonded together in chain-like sequences which may or may not be cross-linked. Some monomers such as ethylene (the related polymer is polyethylene) are simple molecules; others, such as the nucleotides of which DNA and RNA are constructed, are large and complex.

Protoplast: A plant cell from which the cell wall has been removed by mechanical or enzymatic means. Protoplasts can be prepared from primary tissues of most plant organs as well as from cultured plant cells.

Protoplast Fusion: Any induced or spontaneous union between two or more protoplasts to produce a single bi- or multi-nucleate cell. Fusion of nuclei may or may not occur subsequent to the initial protoplast fusion.

Recombinant DNA (r-DNA): A strand of DNA synthesized in the laboratory by splicing together selected parts of DNA strands from different organic species or by adding a selected part to an existing DNA strand.

Regeneration: Developments of whole organisms from single cell cultures.

Restriction Enzyme: An enzyme that cuts and effectively excises a piece of a DNA molecule. Some restriction enzymes cut the DNA at specific points, others appear to cut at random. Restriction enzymes, of which many hundreds have been identified and isolated, are important tools in the excision and transfer of specific gene sequences from one organism's DNA to another's. (See also **endonuclease** and **exonuclease**).

RNA (ribonucleic acid): A polymer of the sugar ribose, phosphate, purine and pyrimidine bases which, as an adjunct to DNA, helps to transmit and implement the genetic instructions for protein synthesis carried on the DNA. Some viruses store their genetic information as RNA not as DNA.

Secondary Metabolites: In addition to the primary products of metabolism, the building materials of which living cells are constructed, plants and animals produce a vast range of secondary metabolites, many of which find application in food, pharmaceutical and other industrial technologies.

Serology: The study of sera (plural of 'serum').

Serological Typing: A technique based upon antibody-antigen reactions by which pathogenic bacteria are identified. It is particularly useful for strains of pathogens difficult to differentiate by morphological methods.

Serum: The watery liquid which separates when animal blood coagulates. Also the blood serum containing antibodies from an animal previously inoculated with a pathogen or pathogenic toxin, used to immunize human or other animals.

Somaclonal Variation: Somatic (vegetative non-sexual) plant cells can be caused to propagate *in vitro* in an appropriate nutrient medium. According to the composition and conditions the cells may proliferate in an undifferentiated (disorganized) pattern to form a callus or in a differentiated (organized) manner to form a plant with a shoot and root. The cells which multiply by division of the parent somatic cells are called somaclones and, theoretically, should be genetically identical with the parent. In fact *in vitro* cell culture of somatic cells, whether from a leaf, a stem, a root, a shoot, or a cotyledon, frequently generates cells significantly different, genetically, from the parent. It appears that during culture the DNA breaks up and is reassembled in different sequences which give rise to plants different in identifiable characters from the parent. Such progeny are called somaclonal variants and provide a useful source of genetic variation.

Somatic Cell: Literally any cell from the 'soma' which includes all cells of an organism except the germ cells. In some instances the term is used to describe undifferentiated cells, such as those found in a cultured callus.

Somatic Embryogenesis: The generation from somatic cell or tissue culture of bipolar embryos, similar to sexually derived embryos. Both sexual and somatic embryos possess a primordial root and shoot.

Somatic Hybridization: The formation of hybrids by fusion of somatic cells, as opposed to the fusion of gametes. The term is commonly applied to fusion of plant protoplasts.

Spore: A reproductive body consisting of one or relatively few plant cells. In a congenial medium a spore may produce a new plant.

Stigma: The female organ in a flowering plant. The enlarged distal end of the style on which pollen alights before passing to the ovule.

Tissue Culture: *In vitro* methods of propagating cells from animal or plant tissue.

Transfer RNA (t-RNA): The molecule that carries amino acids to the ribosomes where an anticodon on the t-RNA reads the codon on the m-RNA and places the relevant amino acid into sequence.

Transformation: The process whereby a piece of foreign DNA is transferred to a cell thus conferring upon it novel characters.

Translation: The process whereby the genetic code present on the m-RNA molecule directs the order of the specific amino acids during protein synthesis.

Transposable Elements: Pieces of DNA that can move from one place to another on one chromosome or move between chromosomes. Sometimes called 'jumping genes'.

Vaccine: A preparation of a pathogenic microorganism or virus, which has been killed or attenuated so as to lose its virulence but which carries antigens. When injected into a living animal the immune system is stimulated to produce antibodies to counteract the antigens. The antibodies remain in the living system thus providing immunity against any subsequent potentially pathogenic infection by the same organism.

Vector: Literally 'a carrier'. In genetic manipulation the vehicle by which DNA is transferred from one cell to another.

Virus: The smallest known type of organism. Viruses cannot reproduce alone but must first infect a living cell and usurp its synthetic and reproductive facilities.

Annex 2:
International Agriculture Research Centres with particular crop or animal production mandates

CIAT — International Centre for Tropical Agriculture: beans, cassava, rice, tropical pastures.

CIMMYT — International Maize and Wheat Improvement Centre: wheat, maize, triticale.

CIP — International Potato Centre: potatoes and sweet potatoes.

ICARDA — International Centre for Agricultural Research in the Dry Areas: barley, lentils, faba beans.

IITA — International Institute of Tropical Agriculture: roots and tubers, cowpea, soya.

ILCA — International Livestock Centre for Africa: cattle, small ruminants, camels.

ILRAD — International Laboratory for Research on Animal Diseases: trypanosomiasis, East Coast fever.

IRRI — International Rice Research Institute: rice.

WARDA — West African Rice Development Association: rice.

The IARCs can capture the economies of scale in research which are unlikely to be available to the smaller ldcs, and their crop-specific mandates permit more intensive work than would otherwise be possible. They hold major germplasm collections for particular crops, and are responsible for the distribution of breeders' material to ldcs. Substantial emphasis is therefore placed on biotechnologies relevant to the production or screening of clonal material derived through *in vitro* micropropagation. These include:

(i) *rapid clonal propagation*: proved successful with banana, tubers and oil palm, and shows promise in fruit, medicinal plants, and forest trees. Its particular advantages include: rapid and disease-free multiplication; the ability to propagate species which are difficult to reproduce vegetatively, and production of plant material on a year-round basis.

(ii) *in vitro conservation*: used extensively with cassava, potato, sweet potato and banana. These techniques offer an alternative to the maintenance of field collections, but problems arise from the genetic instability of crop-specific cultures of somaclonal variants.

(iii) *disease-free plant production*: virus-free tissue can be obtained from over 50 species of plants important to ldcs through the use of chemo- and thermotherapy and antibiotics. Germplasm derived from these is distributed to ldcs to form the basis of their breeding programmes.

(iv) *molecular diagnostics*: faster and more accurate virus detection techniques used on rice, and small grains.

The IARCs are also using the more advanced techniques to facilitate production of germplasm *in vitro*:

(i) *embryo rescue*: facilitates incorporation of genetic material from beyond the primary gene pool into crop improvement programmes. Following sexually-induced hybridization, the embryo is nursed by tissue culture techniques through the early cell divisions until it regenerates as a plant. Using ovule and embryo rescue, together with hormone treatment, scientists have incorporated disease resistance into plants (eg. rice, groundnut and potato) by crossing wild and cultivated varieties which are sexually incompatible.

(ii) *somaclonal variation*: somaclones are screened for tolerance to salt and other environmental stresses. This may prove a useful addition to the variation available to breeders from existing gene pools, providing that *in vitro*-induced variation proves stable in field trials.

(iii) *anther culture*: deriving cultures from pollen cells can shorten breeding cycles and increase selection efficiency. Conventional breeding cycle periods can be cut by more than half, tolerance to cold, drought and salinity can be bred and plant regeneration efficiencies from tissue cultures can be improved.

(iv) *experimental techniques*: more sophisticated and costly techniques include the development of markers designed to identify the large blocks of DNA or chromosome segments that contribute to qualitative characters such as yield. These may assist the breeder in systematically testing and assembling the building blocks of the genomes sought, augmenting the empirical procedures currently used.

Annex 3:
Products of Fermentation

This Annex is an edited extract from IDRC (1985), pp 10-12. Permission to reproduce is gratefully acknowledged.

The products of industrial fermentation fall into several broad classes:

(a) Pharmaceuticals and Health Biologicals

(i) Diagnostics including monoclonal antibodies, other immunoproteins, enzymes, DNA probes for detecting genetic abnormalities in foetuses;
(ii) Prophylactics including antibodies, antigens, vaccines;
(iii) Therapeutics including antibiotics, hormones, steroids and enzyme inhibitors.

(b) Industrial Chemicals

(i) Bulk chemicals including alcohols, organic acids, aldehydes, ketones and simple molecules such as ethylene, methane and peptides from which more complex polymers can be synthesized;
(ii) Fine chemicals such as hormones, enzymes, steroids, polysaccharides and peptide polymers.

(c) Agriculture

(i) Traditional processes include ensilaging and composting;
(ii) Manufactured animal feeds generally contain various antibiotics, amino acids and vitamin supplements;
(iii) Steroidal hormones; recently bovine and porcine growth hormones are reported from recombinant DNA techniques;
(iv) Diagnostics, prophylactics and therapeutics similar to those listed above under 'pharmaceuticals' including monoclonal antibodies, specific antigens, veterinary vaccines;
(v) Microbial pesticides:
these include pathogenic bacteria and toxic substances specifically applicable to the control of known microbial and insect pests and for weed control;
(vi) Rhizobial and mycorrhizal inoculants to improve the uptake of nitrogen and other plant nutrients.

(d) Food

(i) A wide range of traditional fermented foods are produced chiefly at village or small-scale industrial level;

(ii) Industrial fermentations produce an immense range of food additives: nutrients such as amino acids and vitamins; pigments, flavours and flavour enhancers; preservatives; carbohydrate derivatives; modified sugars, synthetic sweeteners; structural and protective colloids, emulsifiers and other texture modifiers;

(iii) Microbial protein from fermented carbohydrates and petroleum by-products and various biological materials.

(e) Other Chemical Applications

In addition to their use in foods, several of the product types mentioned are used in cosmetics, toiletries and non-prescription medicinals. For example the polysaccharide zanthan, produced by *zanthomonas* species, is used to increase the viscosity of food products such as mayonnaise, to a similar end in cosmetics and paints, and as a replacement for cereal starches in the chemical muds of oil wells. Genetic manipulation of microorganisms promises a greater range of macro-molecules for technical use and as starting materials for various chemical conversions.

Bacteria able to convert sulphur compounds are used in increasing the desired metallic content of mined ores of copper and uranium, and in the processing of effluents from pulp and paper manufacture.

The potential use of immobilized bacteria and enzymes for the purification of wastes, effluents and the removal of toxic and noxious substances is greater than yet realised. Other applications foreseen include the removal of unwanted substances from body fluids and the addition to human blood of enzymes, hormones and other essentials in which they are deficient.

Applications

(i) ethanol fermentation from a range of substrates (molasses maize etc)

(ii) methane from biogas digesters

(iii) acetone-butanol fermentations, generally from molasses

Annex 4:
Agricultural Biotechnology in India

With the establishment of the National Biotechnology Board in 1982, India became one of the first developing countries to have an institutional capacity to promote research and training in this new area. The Board's initial objective was to increase among 'line' Ministries and Departments an awareness of the possibilities offered by biotechnology. In February 1986, the Board was upgraded into a department of Biotechnology within the Ministry of Science and Technology, with the mandate to:

i. evolve integrated plans and programmes in biotechnology
ii. identify and finance specific R & D programmes in biotechnology and biotechnology-related manufacturing having a clear end-use
iii. establish infrastructural support at the national level
iv. act on behalf of the government in the import of r-DNA based products, processes and technology
v. prepare safety guidelines for research into and application of biotechnology
vi. initiate scientific and technical efforts relating to biotechnology
vii. promote manpower development in biotechnology
viii. assist in establishing the International Centre for Genetic Engineering and Biotechnology

Among its advisory groups is a Standing Advisory Committee for North America, comprising 7 eminent Indian scientists working in the USA and Canada. The Department is reviewed annually by a joint committee drawn from universities, government and industry. It has an annual budget of approximately 35 m.

In 1987-88, the Department's principal activities included:

Manpower development

Post-graduate and post-doctoral teaching programmes, generally of 2 years' duration, are conducted in 17 institutions. The third batch of students from these programmes is now finding employment in industry and the public sector. The identification of candidates for these courses is highly innovative: applications are considered from a wide range of natural science disciplines (including mathematics), and foundation courses given in biology where necessary. Further institutions are being sought for training in agricultural biotechnology (where only one currently operates) and in marine biotechnology (currently none). Some of these institutions are State Agricultural Universities which have now taken over from the Department the funding of these courses.

Some sixteen short-term training courses for mid-career scientists are conducted annually, a total of over 700 scientists having passed through these by late 1988. The Biotechnology Associateship scheme provides for training at national (4 awards in 1987-88) and foreign levels (31 awards), with a further 17 awards for foreign training designated for the current year.

The Visiting Scientists Programme provides financial and organisational support for foreign scientists to work in Indian institutions on projects of their choice. Technician training was also initiated during 1987-88.

Infrastructural facilities

The Department's strategy is to strengthen existing institutions for teaching, research and development, and industrial processes rather than build new ones. These include:

— germplasm banks for plants, animals, algae and microbes, whose activities in 1987-88 include the provision of 170 cultures of blue-green algae to laboratories, acquisition of over 500 strains of microbe, and computerised data storage and handling; supply of over 30 different animal cell lines to various institutions, and the import of cell lines from the American Type Culture Collection;

— a centralised facility for the duty-free import and supply of fine biochemicals, established jointly by the Department and the Council of Scientific and Industrial Research in 1987;

— establishment in 1987 of a biotechnology information system designed to provide a literature service for (initially) 23 user centres, and to provide scientific data on nucleic acid and protein sequences received from abroad. The computer system will also allow exchange of information on eg. analytical models used in biotechnology research, and will form the basis of a project management information system;

— the commissioning of a 150-litre fermenter;

— the initiation of work on monoclonal antibody production;

— the provision of funding for construction or upgrading of laboratory facilities, such as the Biochemical Engineering Research and Pilot Plant Facility at the Institute of Microbial Technology, Chandigarh.

— the establishment of pilot plants for the tissue culturing of trees suitable for re-afforesting degraded land (aiming eventually to produce 250,000 trees per year), and for an integrated 'start-to-finish' technology for prawn production, with the intention of demonstrating its potential to commercial concerns.

Research and development programmes

Priorities for research to be funded by the Department are determined by Task Forces in specific subject areas, comprising scientists from over 50 organisations in Government, the Universities and industry. Apart from numerous projects in human health, mineral exploitation and industrial fermentation, several agriculture-based projects are being undertaken, including:

— propagation of bamboo by tissue culture;

— immuno-diagnostic kits, embryo transfer and development of vaccines at the National Institute of Immunology. For instance, an embryo donor/recipient herd of over 700 animals has been established and by early 1988, 270 embryo transfers had been made in cows;

— tissue culture in oil palm at the central Plantation Crops Research Institute and the Bhabha Atomic Research Centre. Efforts are also underway to regenerate coconut by tissue culture.

— research on 4 major field crops (rice, wheat, chickpea and *Brassica* species) and, within these, on clearly-identified issues such as sheath blast in rice, karnal bunt in wheat and on the insect genus *Heliothis*.

— production of bio-fertiliser for rice based on blue-green algae.

International collaboration in research and development

A Memorandum of Understanding signed with the USA provides for cooperation across the spectrum of vaccine and immuno-diagnostic methodology, field trails, quality control and vaccine delivery systems. Some 20 Indian scientists are currently undertaking study tours to the USA. A further Memorandum was signed with the USSR in July 1987, and possible areas of collaboration have been discussed with China, west Germany, Czechoslovakia, Indonesia, Vietnam and the Netherlands. The Delhi component of the International centre for Genetic Engineering and Biotechnology has now been established in temporary facilities, but land has been obtained for a new building, staff are being recruited and equipment purchased.

Collaboration with industry

In addition to the participation by representatives of industry in the Department's various advisory committees, links have been promoted through various other means such as the organisation and financing of a study tour by industrialists to the USA in early 1987. A commercial venture for tissue culture of crop plants and ornamentals has been set up in Cochin, and a proposal for export of ornamentals has been cleared by government. Diagnostic kits based on monoclonal antibodies are seen as an area of major potential for industry in both human and animal health, with the possibility of future production of bio-engineered vaccines. Other proposals for the commercial production of hybrid seeds and tissue cultured plants have been made, as have several in non-agricultural areas such as the large-scale production of penicillin.